Praise for *The Change Agents' Handbook*

Traditionally the CEO gets most of the attention and credit for the successful transformation of an organization. This book speaks to the needs of a group of key people whose contribution tends to be overlooked—those who work quietly behind the scenes to support the process.

—David McCamus, former President and CEO
Xerox Canada

A major contribution to the resources available.

—Lynn Cook, Vice President, Operations
University of Alberta Hospitals

David speaks from several perspectives—that of an experienced quality practitioner, a senior executive, and a lead examiner for our national quality award. This reference provides sound advice to the newly appointed change agent, as well as to the other members of the leadership team.

—Douglas Bell, Senior Director, HR/TQM
Petro Canada Resources

This book will make a difference for people who are willing to risk making a difference in their organization. A guide for change agents written by an experienced change agent.

—Duncan MacIntyre, President and C.E.O.
National Quality Institute

Satisfyingly comprehensive! Will be of great value to change agents and also to all *key people, including executives, in a company anxious to improve.*

—Robert G. McGrath
Quality Quest
Former executive with Westinghouse
Commercial Nuclear Fuel Division
(1988 Baldrige Award winner)

Dad, I like it—amusing and straightforward. Sounds good. Great! Fabulous! You used a few words too often though, I think.

—Sarah Hutton, Student

The Change Agents' Handbook

Also available from ASQC Quality Press

Reengineering the Organization: A Step-by-Step Approach to Corporate Revitalization
Jeffrey N. Lowenthal

Process Reengineering: The Key to Achieving Breakthrough Success
Lon Roberts

Incredibly American: Releasing the Heart of Quality
Marilyn R. Zuckerman and Lewis J. Hatala

Team Fitness: A How-To Manual for Building a Winning Work Team
Meg Hartzler and Jane E. Henry

The Way of Strategy
William A. Levinson

The ASQC Total Quality Management Series

TQM: Leadership for the Quality Transformation
Richard S. Johnson

TQM: Management Processes for Quality Operations
Richard S. Johnson

TQM: The Mechanics of Quality Processes
Richard S. Johnson and Lawrence E. Kazense

TQM: Quality Training Practices
Richard S. Johnson

To request a complimentary catalog of publications,
call 800-248-1946.

The Change Agents' Handbook

A Survival Guide for Quality Improvement Champions

David W. Hutton

ASQC Quality Press
Milwaukee, Wisconsin

The Change Agents' Handbook: A Survival Guide for Quality Improvement Champions
David W. Hutton

Library of Congress Cataloging-in-Publication Data
Hutton, David W.
The change agents' handbook: a survival guide for quality improvement champions/David W. Hutton.
p. cm.
Includes bibliographical references and index.
ISBN 0-87389-287-9
1. Organizational change—Management. 2. Total quality management. I. Title.
HD58.8.H88 1994
658.5'62—dc20 94-10361
CIP

10 9 8 7 6 5 4 3

ISBN 0-87389-287-9

Acquisitions Editor: Susan Westergard
Project Editor: Jeanne W. Bohn
Production Editor: Annette Wall
Marketing Administrator: Mark Olson
Set in New Baskerville by Montgomery Media, Inc.
Cover design by Montgomery Media, Inc.
Printed and bound by BookCrafters, Inc.

ASQC Mission: To facilitate continuous improvement and increase customer satisfaction by identifying, communicating, and promoting the use of quality principles, concepts, and technologies; and thereby be recognized throughout the world as the leading authority on, and champion for, quality.

For a free copy of the ASQC Quality Press Publications Catalog, including ASQC membership information, call 800-248-1946.

Printed in the United States of America

Printed on acid-free recycled paper

ASQC
Quality Press
611 East Wisconsin Avenue
Milwaukee, Wisconsin 53202

To my parents, Bill and Mary Hutton

Contents

Exhibits

Foreword

Without an aim in mind and a plan to back it up, trying to improve quality can be like Joe's fishing trip. He starts up the boat, sets up the rod and reel, and away he goes. The first problem is that Joe has never fished before and three miles out he hits a big marlin with a small rod purchased from a corner store with a 10-pound line. The other problem is that the engine just cut out and won't restart, and the auxiliary outboard motor has no propeller. Soon the fact that Joe can't read a compass doesn't matter anymore!

You may never have been fishing or even sat in a boat, but if you work in an organization that is trying to find its way on the quality journey then you may well face the same predicament as Joe. Research tells us that many North American organizations who start a formal quality initiative lose their way or give up within two years, in the process wasting a lot of time, effort, and money. Many household names—institutions as well as companies—have gone down with guns blazing, blaming their own workforce, costs, deregulation, unions, and a host of other reasons for their own failures.

But losing market share, losing the confidence of customers, or going out of business are not punishments from God—they are usually the result of an inability to change and improve the way things get done.

Quality improvement is not a motivation program or a spasm of enthusiasm for some passing fad. It is a system of management that can be learned, understood, and applied. However, as David Hutton outlines in this book, it takes champions to make this happen. Without the dedicated efforts of these change agents, it is "Mission Impossible" for an organization to learn new ways.

These people need help and guidance, so a book for quality champions is a great concept. I know from personal experience that it will help you avoid false starts—not to mention ulcers.

Quality improvement is a journey that is deceptively simple and endlessly complicated. The ideas outlined in this book are not abstract, they are proven and practical and will help you to start out on the right course—or if you're off course and sinking, it will help you to repair the damage and get back on track.

I am pleased that David Hutton has put together this long overdue book for quality champions. It will directly help anyone who is serious about the pursuit of excellence.

John Perry
Vice President
Training and Quality Development
Reimer Express Enterprises, Ltd.

To the Reader

There is nothing more difficult to take in hand, more perilous to conduct, or more uncertain in its success, than to take the lead in the introduction of a new order of things.

—Niccolò Machiavelli, *The Prince*

This book is a personal survival guide for anyone who is striving to create change for the better within an organization. The world would be a poorer and duller place without people who are prepared to take on the established system and make it better. Although they may not always realize it, these individuals are change agents, and they are on a journey that many others have traveled before. This is valuable work that is exciting and immensely rewarding—you can win big in terms of job satisfaction and career.

We know much more about organizations and about creating change than people did in Machiavelli's time, but the task is still challenging. If you do not prepare yourself well, it can be hazardous—to your mental and emotional well-being, to your health, and to your career prospects.

If you are engaged in this type of work, you could use a wise, experienced guide and counselor who can tell you what to expect—someone to help you succeed for the benefit of the organization, and at the same time to look out for your personal well-being. Such a guide could help you answer questions like

- How can we really make things change around here? I know how to organize training classes and roll out new methodologies, but are these actions sufficient?

- What is the role of the officially appointed change agent, and what should others expect of this person?
- How can I tell if we really have any chance of success, given the current status of this organization and the mind-sets of my colleagues?
- Why are so many people resisting what we are trying to do, even though it makes such perfect sense and will benefit everyone?
- Everything went great for the first year, but now we seem to have gone off track. What can I do?
- What aspects of my temperament and outlook will help me to be successful in creating change? What skills and knowledge do I need?
- Why do some colleagues, who I respect, act as if they dislike what we are trying to do?
- I love my work, but I find myself experiencing more ups and downs since we began these changes. This is affecting my family, my friendships, and my health. What should I do?

Most people do not have a close friend with the knowledge and experience to help them with questions like these. *This book tries to be the next-best thing.* True, it cannot help you celebrate your successes, or commiserate over a beer. But, it does offer wisdom and practical guidance distilled from the experience of many people who have already been there. It also shows you how to reach out and use many other sources of help. What you read in these chapters can help you to understand the nature of your journey and to make better decisions along the way.

With the right kind of preparation and guidance, you will be able to do much more than just survive the inevitable ups and downs. You will have the time of your life, broaden your experience and knowledge more than you believed possible, and make a great contribution to the success of your organization.

In order to make things better, many people within the organization, not just the *official* change agent, need to take up the challenge. Whatever your position in the organization, you can choose to work in a proactive fashion to achieve changes for the better.

You will find this guidebook valuable if you fit any of the following descriptions.

- You have been given formal responsibility for creating change, such as overseeing an effort to improve quality or service. You are the official change agent.
- You are the leader of the organization and you are contemplating ambitious changes or have already embarked on them. You need to understand what type of person you should have to help you and what his or her role should be.
- You are a colleague of the formally appointed change agent. You need to know what to expect of this individual, your own role in the change process, and how to implement change within your own area.
- No one is asking you to create change, but you see a real need in your organization, and you intend to do something about it. You are an activist!

The context of the book is an organization-wide change process, based upon the powerful, universal principles of quality management. There are many excellent publications that can help you understand the more technical aspects of quality management, such as the intricate workings of the methodologies, the teachings of the gurus, and what other organizations have done. The appendices provide sources of this type of information, if you need it.

However, no book can cover *everything* you need to know about a quality approach. This book focuses on the *role and the personal needs of change agents.*

If you are striving to create change, then this book is about you—your job, your happiness, your sanity, your future success and fulfillment on the journey ahead.

Read and enjoy!

The champions for change

The advocate

I'm convinced we need to change. We must find ways to convince others, especially top management.

The president

I want to see this organization transformed—with the benefits showing up in our results—or we may not survive.

The official change agent

I want to see this transformation process work really well, and make this a great place for everyone to work. I will learn a lot.

Management

We all want to see improvements, but we're already doing the best we know how. We need to find a better way!

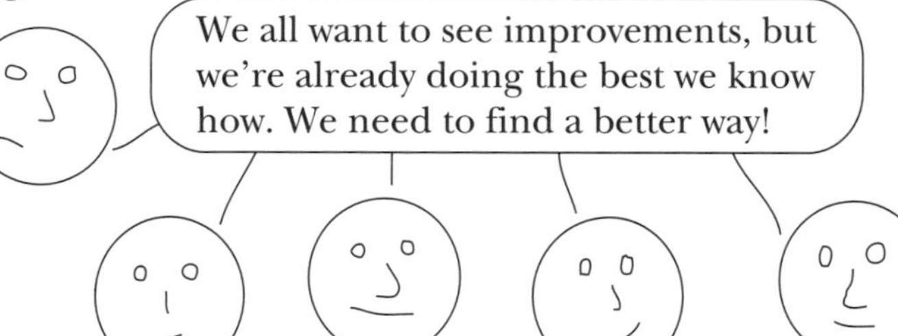

The activist

This place sure needs some changes. I'm not going to sit around waiting for someone else to make the first move.

Acknowledgments

This volume is the outcome of many people's efforts. Numerous colleagues and friends gave generously of their time, to critique the various drafts and to offer valuable suggestions, ideas, and encouragement.

The following people made up my review panel. Through their diligent efforts, a collage of my half-baked ideas was magically transformed into a collection of . . . well . . . fully baked ideas. Doug Bell, David Carlson, Lynn Cook, Bill Delroy, Bob Fisher, Frances Horibé, Roger Gagnon, Fred Leonard, John Long, Jim Ludtke, Karin Lunau, Peter McCulloch, Bob McGrath, Rita Mulroney-Leitch, John Perry, Patricia Taylor, Joan Welsch.

The following people also reviewed material at various stages, offered helpful and thoughtful feedback, and shared ideas for the content. Tony Johnston, Duncan MacIntyre, Leigh Morris, Gino Raffin, Patty Schachter, Greg Watson, Brian Young, Lynda Zavitz.

My thanks also to the six people who took part in the ASQC peer review process; to Gabrielle Bissonnette, my secretary; and to the other committed professionals—Susan Westergard and the rest of the team at ASQC Quality Press—who rolled up their sleeves to help deliver this baby.

I am also grateful to my wife and two daughters who welcomed my pet project, even though it was a particularly demanding and troublesome new addition to the family. Thank you.

Introduction

Quality is never an accident: it is always the result of high intention, sincere effort, intelligent direction and skillful execution. It represents the wise choice of many alternatives.

—Willa A. Foster

This book deals with the role of change agents in orchestrating and supporting a very specific type of organizational change: the adoption of quality management principles and methods as the foundation for running the organization. In order to accomplish this transformation successfully, some people within the organization must understand the following three broad areas: the role of change agents, the science of improving quality, and the process of managing change. This is illustrated in Exhibit I.1.

In this book we will refer to the organization-wide adoption of quality management principles and practices as *a quality approach*. It is often referred to by other names such as *total quality management* or *continuous quality improvement.*

This approach is based upon a body of knowledge—a philosophy and set of principles that are translated into action by means of proven methodologies, tools, and techniques. This approach is not new. It is based upon a line of thought begun by Dr. Walter A. Shewhart of Bell Laboratories in the 1930s, and subsequently developed by teachers and thought-leaders such as Dr. W. Edwards Deming, Dr. J. M. Juran, and Dr. Kaoru Ishikawa.

Between 1950 and 1990, a few pioneering industrial companies, first in Japan and then in the United States, developed these ideas into a practical and comprehensive approach to management, and

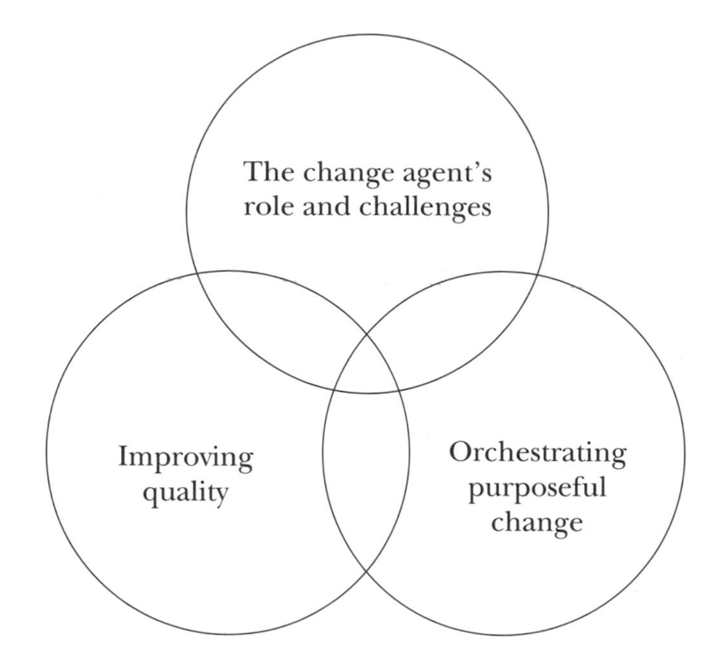

Exhibit I.1 *What this book covers.*

perfected a range of supporting methodologies and tools. Today, this quality revolution is spreading throughout the industrial world, as well as into other sectors such as government, health care and education. Like the Industrial Revolution, this revolution appears to be unstoppable because of the potential it offers for economic and social gain.

Exhibit I.2 sets out the main characteristics of a quality approach, as well as some common misconceptions. If you would like to know more, refer to Appendix A, "Quality in a Nutshell." This is a thumbnail sketch that provides a perspective and context for the rest of the book. It sets out the scope of this field and its historical roots, definitions of quality, and what a quality approach is and is not.

If you are already familiar with the history of the quality revolution, and if Exhibit I.2 tells you nothing you didn't know already, then you are ready to proceed to the next chapter immediately.

Exhibit I.2 *What is a quality approach?*

A quality approach is	A quality approach is not
• led by top management • focused on satisfying customers • designed to involve everyone, and to develop human potential • process oriented • based on a prevention strategy • aimed at continuous improvement • built on cooperative, win-win relationships • aimed at long-term goals • systematic and methodical • based upon management by fact • geared toward public responsibility • a holistic approach	• a project or a program • an add-on • an employee-motivation program • a marketing ploy • a quick fix • a panacea or a guarantee of success • easy to do • dull, mechanical, or boring

1

The Job and the Person

Every man has his vocation. The talent is the call.

—Emerson

CHAPTER CONTENTS

Understanding the requirements of the job and individual, and finding a suitable match.

- Why to appoint a change agent
- What the job is really all about
- Typical scope of the job
- Typical roles
- What the job is not—common misconceptions
- What knowledge and skills you need to succeed
- Why your personal philosophy and attitude are key factors
- Desirable and undesirable characteristics of a change agent
- The magnitude of the task
- How novices can learn on the job
- Check yourself out
- How to recruit an individual suitable for this type of work

This chapter is about the role of an officially appointed change agent, and about the type of person who is likely to perform well in this role. You will find this chapter valuable if

- you already have this type of role, or are considering taking on such a job
- you are looking for a change agent, or you have a colleague working with you in this role and you need to understand what to expect of him or her

The cast of characters

Throughout this book we will talk about a cast of characters who are involved in running the organization and in the change process. Because a quality approach applies equally to any type of organization, from a company or a government department to an educational institution or a charitable body, we will use the following generic terms, which are intended to cover all sectors.

The company, institution, or department will be called the *organization.* The senior person in the organization is the *leader* or *president.* The president and his or her direct reports comprise *top management* or the *top management team.* We will usually talk about the president and senior management as though these were one group of people at the apex of the entire organization. However, the same terms and the same models of change can be applied to any fairly independent entity within a larger organization, such as a division, a business unit, or a branch. If this is your frame of reference, then think of the head of your entity as the president, think of this person and his or her direct reports as top management, and so on.

In addition to these roles, which are found in any organization, we will also refer to the following specific roles, which are peculiar to organizational change, and are described in the literature on managing change.*

---

*Daryl R. Connor, *Managing at the Speed of Change* (New York: Villard Books, 1993), page 282. Copyright © 1993 by Daryl R. Connor. Reprinted by permission of Villard Books, a division of Random House, Inc.

Sponsor
A sponsor for change is someone who has the authority to legitimize the change. The sponsor makes the change a goal for the organization and ensures that resources are assigned to accomplish it.

No major change is possible without committed and suitably placed sponsors. In the case of an organization-wide transformation, the primary sponsors must be the leader and some of the top managers of the organization, or, even better, the entire top management team, acting in concert. With these high-level sponsors signed up, as the change process pervades the organization each manager and supervisor should in turn become persuaded of the need to act as a sponsor for change in their own area. In this way, sponsorship will progressively cascade down through the organization, as illustrated in Exhibit 1.1.

Advocate
An advocate for change is someone who sees a need for change and sets out to initiate the process by convincing suitable sponsors. This is a selling role.

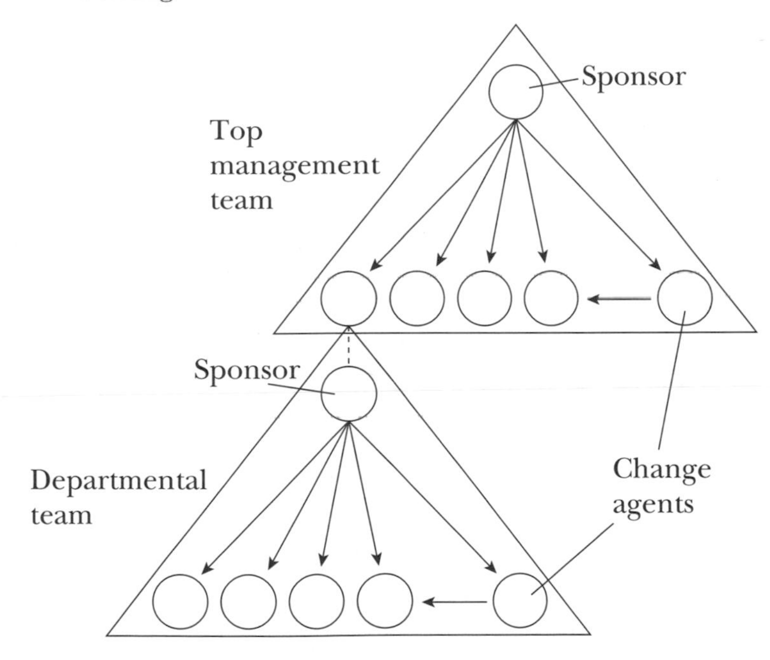

Exhibit 1.1 *Cascading of sponsorship.*

Sponsors are busy people who cannot concern themselves with all the details. They take advice and guidance from others whom they trust and whose judgment they respect. Such trusted colleagues make the most effective advocates.

Advocates need not know in detail *how* the change will be accomplished, and they may not even be involved. For example, the leader may be convinced by a peer in a different organization. An advocate may or may not be powerful in terms of formal position or authority. Their effectiveness is dependent upon the trust and confidence placed in them by those who are in positions of authority.

It is most important never to confuse an advocate with a sponsor. Through the support of advocates, a proposed change may become widely accepted as a great idea or the way to go. But the change is not official—and not likely to go far—until it is approved by the sponsors.

Change Agent

Once the mandate has been granted, a change agent should be formally designated and given primary responsibility for helping management to plan and orchestrate the change.

Although we will often focus on the *officially appointed* change agent within the management team, this book is not written for this individual alone. This book is for anyone who decides to take part in and to support the change process in a proactive manner. From this point on, we will talk about a *Change Agent* when we are referring to an officially designated person, and *change agent* (no capitals) when referring to all the other people who choose to take the lead in this way.

By its very nature, the quality transformation has to be a team effort, but not everyone will start off with the same level of enthusiasm. Some members of the team, besides the Change Agent, have to take the lead and thus encourage others to come along. If you are one of these people, you will not only lead and support change within your own department, but you will also go the extra mile to help others along this path. You may think of yourself as a volunteer change agent—or just a good team player.

Why appoint a Change Agent?

It may be worth addressing the question right away: Does there need to be an officially appointed Change Agent? The answer is a resounding *yes!*

Imagine the following strategy for moving your organization to a new building: You announce the move, knowing that most employees want to be in the new building, and hoping that people will simply make their own way over there. You expect that they will have the common sense to allocate the floor space and offices among themselves, and individually make the necessary arrangements for moving their furniture and hooking up their telephones. With this approach, what is the likelihood of an orderly move? It is similar to the likelihood of an accidental explosion in a print shop creating neatly bound volumes of *Encyclopaedia Britannica.*

Whether we are looking at an office move or major organizational change, a similar principle applies: Unless someone takes responsibility for masterminding the whole affair, either nothing will happen, or there will be complete chaos. In fact, even with someone orchestrating the changes, the process rarely takes place without incident.

Perhaps the president feels that he or she should remain personally responsible for an initiative that is so important and so all-embracing. This is absolutely correct, but he or she still cannot do everything. The president needs others to play their appropriate parts. Every team member must take part in developing a plan, take ownership of their piece of the action, and deliver on their commitments. The president also needs many people within the organization to embrace the change and to show leadership.

None of this will happen unless there is a Change Agent assigned, to guide the team as they develop the plan, to help figure out how to accomplish the tricky or technical bits, and to provide ongoing support for those who pioneer the changes. So, it is unwise to attempt major change without a formally designated Change Agent. In fact, experience indicates that the skill and dedication of this individual is a key success factor.[1]

THE JOB

What is the job all about?

The Change Agent is someone whose role is to support the president and the top management team in bringing about a purposeful transformation of the organization. This transformation involves

- Helping people to change the way they think, and what they think about, in performing their jobs (for example, the way they look at problems, errors, and waste; their attitudes toward each other, the people who report to them, and suppliers and customers)
- Changing the norms (for example, accepted standards of behavior toward colleagues, daily work practices, how meetings are run, acceptable standards of service to customers)
- Changing the organization's systems and processes, to make it better able to reach its goals (for example, management processes such as planning and budgeting, information systems, customer and employee feedback systems, as well as the key operational processes used to create and deliver products and services. This also includes establishing measurement systems that enable the organization to understand how it is improving)

These tasks include some tangible, "hard" objectives, and others that seem intangible or "soft."

There is a discipline and rigor involved in tasks like building relationships and changing behavior, and these aspects of the management system also can be measured and improved. These soft issues are just as essential to success as the more tangible, or hard, issues—and they often require more time and effort to deal with successfully.

This is an important observation. *Like the change process itself, the Change Agent's job has both hard and soft dimensions.* Both *are critical to success.*

The scope of the job

Let's get down to specifics regarding what the Change Agent actually does on a day-to-day basis. The following are typical tasks.

The Change Agent educates and enlightens the top management team, and orchestrates its improvement-related activities. For example, the Change Agent

- Proposes and organizes educational activities, such as formal workshops, guest speakers, "lunch and learn," visits to other organizations, and so on
- Organizes an assessment of the status of the organization to identify strengths and shortcomings in the current approach to managing quality
- Orchestrates the development of a plan for improving the quality to help the organization meet its goals and ensures the integration of this improvement plan into the business plan
- Helps organize the meeting schedule of the senior management team to ensure that time is allocated to planning for improvement, and for monitoring progress in implementing the plan

The Change Agent supports and advises the leader and other colleagues as a senior manager, as a subject matter expert, and as a helper. For example the Change Agent

- Plays an active part, like any other member of the team, in discussions and decision making regarding general issues that affect the organization—from the development of business strategy to the implementation of the no-smoking policy.
- Provides guidance to the team on technical issues having to do with quality, such as selecting suitable strategies and methods, choosing suitable individuals as quality coordinators and trainers, identifying suitable suppliers for training materials or consulting support, and helping management to monitor the health of early projects.

- Acts as a considerate and discreet advisor or provides a second opinion on behavioral issues, relationships, and personnel issues. For example, this person should be able to coach the leader on his or her management style, giving positive feedback as well as pointing out if the leader sometimes falls short as a role model. This requires judgment, tact, and a good sense of timing, as well as a leader who is open to such feedback.

  If the Change Agent is very capable in the role of coach and confidant, he or she may be asked to coach colleagues confidentially, perhaps even in basic skills like planning or giving presentations. This does happen, but only when a high level of mutual trust has been developed—when senior people do not want others to know that they feel they are lacking such skills.
- Challenges the top management team when the evidence suggests that the process, or the team's commitment, is faltering.

The Change Agent manages specific projects. Like any member of the senior executive team, the Change Agent will act as project manager or project sponsor for specific initiatives. Usually these projects will be directly related to the quality improvement efforts, such as setting up a regular customer satisfaction survey, or deploying quality-related training throughout the organization.

The Change Agent develops and manages a support network. In an organization of more than 100–200 people, there usually will be others assigned to support the process in addition to the top management Change Agent. Typically these people undertake tasks such as training, facilitation, and coaching teams—often on a part-time basis. They undertake these tasks as part of, or in addition to, their normal jobs.

These people make up the internal support network that supports the change process. They need to be identified, appointed, trained, developed, and supported. This network is essential to the success of the improvement efforts, and the Change Agent must lead and develop it, even though the individuals involved usually report directly to other managers.

Possible roles of the Change Agent

Exhibit 1.2 lists roles that are appropriate and inappropriate for the Change Agent.

Like any other manager, the Change Agent may play a wide variety of roles in the course of a day, switching between them according to what is required at the time.

Some of these roles are described as inappropriate for the following reasons.

- *Commander Quality*. Change agents are support people; it is not their role to hand out instructions. Rather, it is line management's function to allocate responsibility and accountability. Even if the leader says to the Change Agent "I want you to go and *make* them do it," attempts by a staff person to command action are usually doomed to failure. If a line manager asks a staff person to do something like this, an appropriate response might be "Gladly—if you give me your job."
- *Progress chaser*. The Change Agent may monitor progress in some areas, but progress chasing (that is, chasing after others, trying to persuade them to take action) is an inappropriate role. Rather, the Change Agent should work with line managers to establish reliable systems for measuring and tracking progress.

Line managers are responsible for improvement and should use these systems to report on their own progress to their manager

Exhibit 1.2 *Appropriate and inappropriate roles.*

Appropriate	Inappropriate
• Visionary • Advocate • Navigator • Confidant, supporter • Coach • Subject-matter expert • Role model	• Gopher • Commander Quality • Spy • Progress chaser • Self-promoter or credit-taker

and peers. The most effective process for monitoring and chasing progress is regular review by the leader and the management team.

• *Spy*. Integrity is the Change Agent's most valuable asset. It is the foundation for open and honest relationships with peers, and it allows the Change Agent to earn their trust. This will be impossible to achieve if the change agent is perceived as the leader's spy. It is important to avoid any arrangement or behavior that could be misinterpreted in this way.

Top management often assumes that the head of a staff function, such as marketing, human resources, or quality, should be able to make the improvement process happen by taking on some slight extension of his or her current duties. This is a natural, but dangerous, assumption. Such individuals already have a full set of responsibilities, which are significantly different from those of the Change Agent. These current responsibilities may not mix well with the new ones, and there are other issues to consider, such as skill sets, aptitude, and work load. Similar issues also arise when a senior line manager is assigned the additional role of Change Agent.

This type of dual responsibility is not a no-no—such an arrangement can be made to work well. However, it is definitely a red flag, which calls for caution and an understanding of the pros and cons. This is a very important issue. If such an arrangement is being considered in your organization, you should refer to chapter 7, "Internal Resources," for a more detailed discussion about the Change Agent's base of operations.

THE PERSON

Integrity without knowledge is weak and useless, and knowledge without integrity is dangerous and dreadful.

—Samuel Johnson

There is no single formula for a successful Change Agent. They come in all shapes and sizes, and they have different personalities and approaches to the job. No one is equally good at every aspect or knows everything that might be useful in doing the job.

However, there are some personal characteristics that seem to be essential, and there are some key areas of knowledge, skill, and experience. Let's start by looking at the characteristics that the leader should be looking for in a candidate. These fall into three main areas.

1. Personal skills and attributes
2. Knowledge and experience of the business
3. Knowledge and experience of quality

Many of these are also characteristics that will help anyone—the leader, managers, or frontline employees—to participate in and support the change process.

Personal skills and attributes

The chosen person should

- Believe in the job. In order to be effective, the Change Agent must identify with the aims of the change process, the underlying values, and the cultural changes sought.
- Be effective in a senior management role if he or she is operating at this level of the organization. This person must be able to deal with the type of issues his or her peers deal with (for example, more strategic than operational), and be familiar with the way senior management thinks and works. If the individual is lacking experience within this peer group, he or she must have the ability to step up to it quickly.
- Demonstrate integrity and the ability to earn the trust and respect of peers—as well as other people in the organization
- Be a team player. Improving quality is a collective undertaking, not just a personal project. This individual should have a natural inclination to help others, involve others, and collaborate with others at every opportunity
- Be competent in basic management skills, such as communication and interpersonal skills, working with groups, negotiating,

planning and budgeting, project management, and problem solving. Because this person will be a role model, the skills required for a participative management style are particularly important. These include listening, soliciting feedback, coaching, encouraging diversity, and managing conflict.
- Have patience, persistence, and a sense of humor. These qualities might be summed up as *resilience.* They are not just nice to have—they are essential!

Knowledge and experience of the business

The Change Agent needs to be able to work with people throughout the organization in their own language and understand the kinds of problems they face. He or she should have some understanding of the products or services, the marketplace, and typical customers or clients. He or she should understand the formal organization structure and the informal networks and alliances. He or she should know the key players personally—what they stand for, who carries clout, and who influences opinion. In short, the Change Agent needs

- Familiarity with this type of business or institution
- Familiarity with this specific organization, its situation, and its people

Knowledge and experience of quality

This person needs to thoroughly understand quality management—which is a large field—or be willing and able to learn quickly. The knowledge required is primarily strategic, not nuts and bolts.* For example, the change agent does not need to be able to do things like

- Teach people about the inner workings of design of experiments, statistical process control, or quality function deployment.

---

*The methodologies and tools mentioned on the next few pages are described in Appendix C.

- Conduct a benchmarking study at the drop of a hat.
- Determine confidence levels in an analysis of customer survey data.

However, this individual should, in time, learn to

- Teach other people (especially peers) about quality, both formally and informally, in terms that make sense to them, and without making them feel ignorant or uncomfortable.
- Explain how methodologies such as quality function deployment or benchmarking may, or may not, be useful to the organization.
- Act as a role model, not just in management style, but in applying the basic methods and techniques (like formal problem solving, using the seven quality control tools, conducting a process improvement project on a key management process, or applying some of the seven management tools in appropriate situations).
- Respond convincingly and accurately (no bluffing) to most of the questions that people may bring up in their efforts to understand quality improvement, or in their efforts to demonstrate that this quality improvement stuff is silly.
- Help the management team to understand the current status of the organization in terms of how well it is organized to improve quality, service, and productivity, and thus achieve its goals.
- Guide the management team in the development of strategies and plans that are appropriate to the organization's current status and needs.

Why attitude is key

We can always acquire new knowledge and learn new skills to equip ourselves for a new job. However, there are personal aspects that we cannot easily change—and probably don't want to. These are our natural aptitudes, our personal preferences, our beliefs about the world. Many of these boil down to *attitude.*

Attitude is key in this job because it is not easily changed, and because it influences

- The Change Agent's credibility and acceptability to others
- The Change Agent's ability to perform some key tasks and to persist in the face of setbacks

Here are a few examples to illustrate the importance of your own attitude.

Level of comfort with a participative management style

What management style do you feel is appropriate for a senior person? Do you see anything to be gained (or learned) from personal contact with frontline people? When you are with lower-ranking people, do you expect to be doing the talking while they listen? If they challenge your opinions or offer information that contradicts your views, do you find this disrespectful or inappropriate? How would you feel if they went off and solved some major problem on their own, without asking your permission or involving you?

If the reins of power are very important to you, that is no crime—but you may find it hard to share more information and let others make decisions more often. If you need the institutionalized respect that comes with rank, you may never get close enough to employees to learn how to engage their full talent and enthusiasm.

Yet, it is part of the Change Agent's job to be a role model for more effective management styles. No one can be effective advocating one set of standards and living another. You cannot be credible as an advocate for participative management training while administering the training program like Attila the Hun. You cannot help to bring about a more open and trusting environment if you act in a secretive or devious fashion.

Patience

If you are a very impatient person who likes to see quick tangible results and move on to the next task, then this job will surely drive you to distraction. You may do a poor job for this reason alone. Organizational change takes time, no matter how many resources are available, no matter how well-designed the change process, or how well-chosen the methodologies.

If your personal preference is to experience the instant gratification of white-water rafting, why would you take up gardening, where it takes months or even years to see the results of your efforts? You will find yourself checking progress in your garden with a stopwatch instead of a calendar, and pulling up the plants to see if the roots are growing. (See Exhibit 1.3 for a related anecdote.)

Belief in the Job

Another attribute that cannot easily be acquired like putting on a uniform is a strong belief in the value of the job. This belief stems from personal identification with the aims of the change process and the underlying principles and values. Given time and experience, the Change Agent's instinctive affinity for the task will develop into a personal vision of the future and a strong belief that this vision is possible and worth striving for.

Change agents need this vision to enable them to engage the support of others and to sustain their own personal commitment and enthusiasm. This vision helps a person to keep going when the going gets tough.

The desired attributes of a change agent are listed in Exhibit 1.4. Many of these are no more than one would ask of a good all-around manager. You may say "Any manager worth his or her salt will behave in this way." However, the reality is that, although many people would like to behave like this, if only for their own self-respect, the traditional management system discourages many natural, positive behaviors.

Exhibit 1.3 *Combining patience with a sense of urgency.*

There is a story of a French general in retirement who decided to plant an avenue of beech trees on either side of the long avenue leading up to his chateau—for the benefit of future generations of his family.

He gave the head gardener instructions, and a few days later inquired about progress. The gardener admitted that he had not yet planted the trees, adding "Of course, you realize that they will take several hundred years to mature."

"Exactly!" responded the general, "there is no time to be lost."

Exhibit 1.4 *Desirable and undesirable characteristics of a change agent.*

Desirable	Undesirable
• Believes in the approach (it is "the right thing to do") • Patient and persistent • Honest, trustworthy, reliable, demonstrates integrity • Positive and enthusiastic • Confident, but not arrogant—able to admit ignorance and ask for help • A good listener, observant of others' feelings and behavior • Flexible and resourceful in searching out alternative ways to achieve goals • Not easily intimidated • Able to maintain perspective (good sense of humor) • Willing to take personal risks and to take on a challenge • Able to recognize and deal with office politics without becoming involved in the politics • Inclined to adopt a cooperative, inclusive style	• Interested in the job only as a stepping stone (no vocational interest) • Highly impatient, lacks persistence • Unreliable, devious, or untrustworthy • Unenthusiastic, or unable to convey enthusiasm • Has a strong need for praise or recognition from others • A poor listener, insensitive to others' feelings • Autocratic, arrogant, cold, unapproachable, inflexible • Very status-conscious • Moral putty (for example, readily adapts values/beliefs to suit the circumstances) • Disorganized (for example, struggles to manage small projects) • Highly averse to personal risk (for example, to image, career) • Highly political or manipulative • Inclined to adopt an adversarial or secretive style

Bitter experience has taught many people that in a work setting you cannot trust anyone. You must constantly promote your own personal interests ahead of others. You must cheat the system, if you can, to gain an edge. They believe that you would be a fool to do any differently, because everyone else is acting like this. Many managers have been consistently rewarded for building up empires and protecting their turf; for making themselves look good by altering or concealing the truth; for spectacular and highly visible fire fighting rather than painstaking and low-key work to prevent problems from arising.

Most people are capable of acting quite differently if the dynamics of the workplace are changed and different behaviors are rewarded. However, nothing will change unless there are some role models—people who are prepared take the lead in setting the new norms.

To be credible and effective, the Change Agent must be one of these role models. This requires a personal attachment to the values that underpin these behaviors and a willingness to take a risk and be out of step with others. The Change Agent should not just be someone who can demonstrate these attributes, but someone who can be relied on to do so, even when under strong pressure to act differently.

The magnitude of the task

When we discussed the Change Agent's knowledge and experience of quality, one of the key requirements was the ability to provide strategic guidance. This is critical to the long-term success of the change process, but it requires a depth of knowledge that the newly appointed Change Agent rarely possesses.

In this section we will illustrate this point to underscore the magnitude of the Change Agent's task and this individual's need to draw upon various sources of help. We will also explain why this need not become a problem.

In order to become really good at this job, a Change Agent needs to learn a great deal about quality at a strategic level. This involves

- A grasp of systems thinking. In a quality approach, strategic planning is guided by an understanding of the organization as a system to be optimized.
- An understanding of what strategies and methodologies are available, what these are good for, how they interrelate, which ones need to come before others, and why.
- An understanding of strategic planning.

The Change Agent will often be confronted with strategic issues regarding methodology. This will happen more often as colleagues become interested and learn more themselves. Here are some examples of the kind of guidance the Change Agent should learn to provide. The Change Agent should learn to

- Explain to the team why formal process benchmarking might not be helpful until management understands its own

processes. The Change Agent should also learn to suggest better ways of gleaning ideas and information from other organizations during the early stages.
- Harness an interest in one approach such as ISO 9000, process reengineering, or empowerment; put these approaches in context; and show the team how to apply them as components of a broader strategy that will help the organization to meet its goals.
- Talk to operations people sensibly about the relationship between design of experiments and statistical process control, and to the design people about the strengths and limitations of quality function deployment, in order to help them determine their local priorities.
- Explain how a Hoshin planning system differs from the company's current management by objectives system, and help the team figure out when in their quality journey the organization might begin to explore such an approach to planning.

The Change Agent should also learn to

- Help the team figure out how to wring some convincing tangible benefits out of the system relatively quickly—in order to renew the mandate and as insurance against normal setbacks.
- Decide what kinds of strategies, tools and techniques may particularly appeal to a given function (for example, design, marketing, or human resources) to stimulate interest and get its people involved, while at the same time maintaining consistency with the overall direction.
- Figure out how to build on any preexisting quality improvement efforts that may be going on in different parts of the organization, to integrate these into future plans, and to cause them to converge rather than to diverge further and possibly conflict.

Walking on water

The list of learning requirements for the ideal Change Agent may seem daunting, but it is not Utopian. Guiding this transformation is not a simple or easy task. No one would expect an accountant, however bright, to become a top-notch engineer in a year or two—or expect a marketing specialist to become an expert lawyer simply by attending some conferences and reading a few books. It is equally unrealistic to expect anyone to acquire a thorough understanding of a quality approach in a short time. *There is a great deal to learn, and so a major investment of time, energy, and dedication is required to become an accomplished and effective quality change agent.*

This expertise is key to success in the long term. Without such expertise, this person, who is entrusted with the task of guiding the organization, is like a navigator who cannot yet read the map, let alone help *create* the map. Yet, most organizations start out on this journey with Change Agents who, although experienced and capable managers, are novices in this field. What are the chances of success, given this common starting point?

Fortunately, there are many ways to compensate for lack of experience; for example, by parceling out the learning tasks and by judiciously drawing upon external resources. By using such methods, the novice Change Agent can meet the important needs of the team, while at the same time learning on the job. So, the Change Agent does not have to walk on water, or know everything there is to know about quality, life, and the universe.

Learning on the job

There are several reasons why the Change Agent and the management team will succeed if they persist, even though they may start with little knowledge or experience of this approach.

- This is a team effort. Everyone on the management team has to learn enough to create improvement within their own areas of responsibility. The ability of the team to create improvement is determined by what the most enthusiastic line managers can accomplish in practice—not by what the Change Agent can visualize in his or her head.
- The team has to learn from firsthand experience. Following the guidance of an in-house expert is fine up to a point; but the team has to make the final decisions, experiment a bit, and make some mistakes. This is part of the learning process. The Change

Agent only needs to head off decisions that might cause fatalities rather than bruises.

• The process can, and usually should, be started in small ways, using simple methods, while management gains an understanding of and a feel for the methods.

• There is usually no compelling reason to make instant decisions. When you encounter issues where you feel out of your depth, you often can take more time to explore, to gather more information, and to consult others.

• At any point in time, the management team only needs to know enough to take the next steps toward achieving its goals. For example, if a critical goal is to improve responsiveness, you don't need to understand all the subtleties of process management. Perhaps all you need to do at first is master a simple approach to reducing cycle times, and/or focus problem-solving efforts on time-related issues.

• Your customers and your employees can tell you so much today about how to improve your organization. You can make a good start just by using this input and employing some basic problem-solving methods, even if you do not yet have a very clear understanding of the big picture, or have not yet figured out your longer-term intentions.

• You are not the first. There are excellent external resources you can draw upon to help you, including peers in other organizations, consultants with extensive implementation experience, and associations and other organizations that can provide all kinds of information.

Chapter 8, "External Resources," outlines such external resources and explains how to make use of them. No matter how experienced we become as change agents, we all need these resources—sometimes to fill a gap in our own personal knowledge and experience, or to help us figure out some problem, but always to learn more.

• There is no better alternative. Outside experts can bring you their experience, but ultimately only the people within the organization know enough about their own situation to make wise decisions.

So, the novice Change Agent, supporting a management team that is learning about quality improvement for the first time, will be successful by

- Ensuring that the entire change process, including the learning, is a team effort.
- Focusing learning efforts on areas that are most relevant to the organization's goals.
- Recognizing the boundaries of his or her own knowledge, and learning how to access and make full use of other sources of expertise as required.
- Learning fast enough to meet the needs of other people in the organization as they become more advanced in their thinking. Fast learning is also essential in order to retain credibility (and, hence, influence and usefulness) as a navigator.

In the long term, the organization should aim to develop a high level of internal expertise in quality, shared among many people. The Change Agent needs to lead this learning process and to teach and coach others. Towering expertise in quality is not essential from day one. But a willingness to learn, and the wisdom and humility to recognize how much there is to learn, are important attributes for any change agent.

A self-assessment checklist

If you are a prospective change agent, the table in Exhibit 1.5 provides a checklist to help you examine your own particular strengths and challenges. Don't take this too seriously—clearly it is not a scientific assessment instrument, and the suggestions for action might seem simplistic. It is designed only to give you some ideas.

In Chapter 9, "Personal Survival and Growth," we will explore in more depth what it feels like to do this work; what are the rewards, the personal challenges, and ways of dealing with these. If you are having real difficulty in deciding whether this type of work is for you, then Chapter 9 will help.

Exhibit 1.5 *Check yourself out.*

Desirable attribute	Symptoms of a weakness	Likely consequences if not addressed	Possible actions
Patience	You develop physical symptoms (for example, palpitations, high blood pressure) when you have to wait in line. You tend to finish other people's sentences for them. You eat breakfast at night to save time in the morning.	You may begin to lose heart or lose interest before the process has had time to produce results; or find yourself wondering what's gone wrong and looking at alternative approaches.	Determine feasible time frames and milestones in advance. Identify smaller and more frequent intermediate milestones, including observable changes in behaviors. Keep a log of all achievements, and celebrate every success.
Persistence	You get angry or discouraged when things don't work out exactly as planned. You often feel that events are conspiring against you.	You may feel thwarted and unnecessarily discouraged when you should simply be biding your time and waiting for a better opportunity.	Hold on to the desired goal by being flexible about the means. When you seem to be up against a roadblock, go back to first principles, reexamine what you are really trying to accomplish and brainstorm alternative approaches.
Belief in the process	You find yourself agreeing (or stuck for a response) when colleagues put forward standard objections such as "This is just motherhood" or "It's a nice theory, but not practical in the real world" or "We can't afford it."	You may not be a very convincing advocate, and your lack of strong, thought-out responses may undermine your credibility.	Build your information network. Spend time with people who have made quality improvement work in their organizations. List the comments and objections that throw you, think about them, and discuss them with other change agents.

Exhibit 1.5 *(continued)*

Desirable attribute	Symptoms of a weakness	Likely consequences if not addressed	Possible actions
Willingness to give away the credit and recognition	You become upset when you feel that others are angling for the credit for some success or are underestimating your contribution. You are inclined to toot your horn a lot. When your child receives some award or acclamation, you want everyone to know that you are the parent.	Your job is to quietly ensure that others are recognized for their efforts. If you believe subconsciously that there is only a fixed amount of credit to go around, and you have a strong need to get *your share,* this will become a divisive issue in your relationships with others. Everyone will end up feeling resentful and cheated.	Think about whether your first choice is to see some changes take place or to be a hero. If you struggle with the idea of foregoing personal recognition in order to get the job done, this probably isn't the job for you.
Reliability	Colleagues sometimes don't take your promises seriously. You are often late for meetings and other appointments. You have developed the excuse into an art form. Your excuses to others are varied, plausible, witty, and frequent. Friends lie to you about when the movie or the dinner starts in order to get you there on time.	You may be a poor role model sending the message that it doesn't matter if you don't keep promises to others. You may have difficulty in meeting commitments that are important to the change process.	Tackle this like stopping smoking. First, look for ways to avoid self-deception on this issue; for example, keep track of broken commitments, tell some close colleagues of your desire to improve, and solicit ongoing feedback. Then search for root causes, such as reluctance to say no, lack of prioritization, poor time management, and so on.

Exhibit 1.5 *(continued)*

Desirable attribute	Symptoms of a weakness	Likely consequences if not addressed	Possible actions
Sufficiently confident to admit ignorance and ask for help when appropriate	When someone uses some jargon you don't understand but think you should, you nod your head wisely rather than ask for an explanation. You prefer to drive in circles for hours rather than ask someone for directions.	You may commit serious errors of judgment by failing to recognize the limitations of your knowledge or by discounting other valid points of view. You may lead your colleagues (and the organization) in the wrong direction.	Lack of self-confidence is like a disease, which can seriously undermine your abilities. Here are some things that might help: place more trust in other people's positive opinions of you; join Toastmasters and master public speaking; make an inventory of your intrinsic good points; discuss your self-doubts with someone you trust.
Able to convey enthusiasm	People often nod off during your presentations. People sometimes think you're depressed when you're actually on top of the world. Friends fall asleep while you are telling them about your most exciting experience.	You may find it hard to get others as excited as you are about improving quality.	Work on your communication skills. Work on identifying other people's work-related interests and concerns, and link your message to these issues. In this way, they are more likely to become enthusiastic even if your demeanor is a bit flat.

Exhibit 1.5 *(continued)*

Desirable attribute	Symptoms of a weakness	Likely consequences if not addressed	Possible actions
A good listener, observant of others' feelings, behavior	Other people often become difficult—flying into a rage or getting upset without any apparent warning. You don't find other people's personal concerns interesting— you would have difficulty in writing out an inventory of your colleagues' personal interests, family status, personal problems, and so on.	The big challenges in managing change are not the technical ones, but those related to human relationships and emotions. If you have a blind spot in this area, you may be blissfully unaware of the hot issues boiling beneath the rational surface. Neglect of these issues, or insensitivity toward them, is a sure recipe for failure.	This won't be easy. To achieve a change in your style will require time and unrelenting effort on your part; but, it's well worth it in the end. You could start with training on listening skills. You could play observation games with your partner. Most people have the ability to tell at a glance what mood another person is in, but some people don't realize that they have this ability, or don't trust their instincts. You could work harder at considering other people's feelings at all times.

Exhibit 1.5 *(continued)*

Desirable attribute	Symptoms of a weakness	Likely consequences if not addressed	Possible actions
Not easily intimidated	You really don't like being out of favor with your boss or your colleagues, or going against the flow. When this happens, you begin to feel insecure, isolated or guilty. Your opinions and your behavior are easily influenced by pressure from others. You always obey people in authority and feel slightly nervous in the presence of parking attendants and bus conductors.	You need to hold onto a minority view on many issues, ranging from the extent to which decision making should be shared with employees, to the extent to which it is realistic to put the customers' needs first. If you are too easily influenced by others, you may, like them, succumb to short-term narrowly focused pressures and lose sight of your goal.	Your task is to hang onto your beliefs and to find ways of making a quality approach useful to others. Reinforce your confidence and understanding of the process by contact with other more experienced change agents. Work to ensure that the team jointly discovers unwelcome facts and new perspectives, rather than always being the messenger yourself.

Exhibit 1.5 *(continued)*

Desirable attribute	Symptoms of a weakness	Likely consequences if not addressed	Possible actions
Able to maintain perspective (for example, good sense of humor)	You tend to focus 100 percent of your energy on the task in hand. You find others much less concerned than you are about the future of humanity and of the universe. You are upset for weeks if your work project suffers a setback or if you receive a disappointing performance appraisal.	You may suffer excessive stress and anxiety during the inevitable ups and downs which take place as the change process unfolds. Your ability to influence others may suffer, especially if your focus narrows to exclude their concerns, or if your seriousness about the task is perceived as being obsessive.	Don't take on responsibility for more than you should—the improvement process is a team effort, not your sole responsibility. Draw upon your friends to help you blow off steam and put things in perspective. Watch yourself for signs of burnout.

RECRUITING THE RIGHT PERSON

This section touches briefly on the task of finding suitable people to support change. This is a task that the management team faces once the decision has been made to designate a Change Agent. It is also a task that the Change Agent may face in identifying others to support the process within the organization.

Recruitment options

The following are typical recruitment options that may be considered.

- Lateral transfer of one of the management team members into a new position, giving up all or most of his or her previous responsibilities
- Assignment of one of the management team members to the task as an additional responsibility (perhaps with additional full-time direct reports to assist)
- Elevation of a promising individual into a new position within the management team
- Recruitment of an outsider

Issues to consider

When the process is first being launched, it is unusual (except perhaps in large organizations) to find someone internally who already has the desired level of quality-related knowledge and experience. How would they have acquired it? If no ideal internal candidate can be found, the options available may be to recruit an outside quality expert, or to develop someone from within the organization into this role. There are many benefits to choosing someone from within the organization.

- Internal candidates' characteristics and abilities are known quantities—you don't have to guess or take a risk based on interviews and references.

- The individual already knows the organization, the type of business in which it operates, the people, the politics, and so on.
- The individual is already known and respected by people within the organization and already has working relationships established with many key people.

An outsider is a much higher risk in these areas. Even if their personal attributes are ideally suited to the job, it will take a long time for them to become established and accepted. Also, you can encourage a capable and suitable person to learn all about quality, but you cannot do much with someone who knows quality but proves to be antagonistic, unreliable, or untrustworthy. You might even be unfortunate enough to recruit someone whose quality-related expertise proves to be dubious. How well qualified are you and your people to evaluate a candidate's quality-related credentials? Many organizations therefore choose, for good reasons, to develop one of their own up-and-coming people into a quality expert.

There are also some very poor reasons for taking this route. The following ideas are dangerously wrong.

- "How much can there be to learn about quality? Surely a couple of classes and a few months on the job should do it."
- "There are some people around in the organization that we could spare right now."
- "We don't want to go to the expense of hiring someone and then find that this quality stuff doesn't work."

Would managers adopt such attitudes if this individual was going to perform brain surgery on some of them? Yet the management team is contemplating changes to the organization that may be just as significant and which require a similar level of skill in execution.

The recruitment process

No matter how the recruitment process is performed, there are some important outcomes that should be achieved.

Exhibit 1.6 *Passing through quality.*

It is a fact that people often briefly pass through this position, especially when it is part of the top management team. The job is used as a development opportunity that soon leads to something else. Here are some implications of this strategy.

- Working in this role is an accelerated (and forced) learning experience. It may serve as a training ground to develop more champions for the change process and to equip senior people with a valuable set of new skills.
- Because the people following this path still have to climb up the learning curve and may then be replaced, they may need access to an experienced mentor—a kind of "shadow" change agent.
- This position may have some symbolic significance within the organization, and it may be unsettling to employees to observe frequent changes of the incumbent. Such changes may reinforce a sense of "flavor of the month," or create the idea that the job is a penalty box. If the plan is to assign people to the position for a term only, then it may be prudent to announce this up front.
- Succession planning becomes more important, so that the next incumbent prepares him or herself and is ready to step in at short notice.

Using this position as a training ground may be a good strategy once there is some depth of knowledge and experience in the organization. However, during the early stages, it makes more sense to ensure that the change agent is someone with a long-term commitment to the task, rather than risk the job being treated merely as an interesting learning experience or a stepping stone to something else.

- The people involved in making the choice should be dealing with a known quantity. There should be a high level of confidence that the experience, capabilities, and attitudes of the chosen candidate have been accurately assessed.
- The leader should be completely satisfied with the choice of Change Agent.
- The rest of the management team members should also have a say and agree with the choice.

2

The Situation

Reconnaissance is never wasted effort.

—A military saying

CHAPTER CONTENTS

How to assess whether the organization is ready for change and whether success is possible.

- Why to assess the situation
- The *must haves*
- The change agent's wish list
- Dubious blessings
- Potential major barriers
- How to investigate
- Arriving at a decision
- What to do if the commitment to change is lacking
- How to build management commitment to change

No sensible person would willingly take on a major new task without trying to determine what he or she would be getting into and the chances of success. Often, the difficult part is knowing what to look for and what questions to ask. In this chapter, we discuss how to size up the situation—how to find out what this organization has going for it and what barriers to change exist.

We will assume that the individual conducting this investigation is the prospective Change Agent. However, whether you are the future Change Agent, the leader, or someone else who wants to see some changes, it is wise to start by determining the size of the challenge ahead.

Why to assess the situation

Change is never easy, but sometimes it is simply out of reach because of circumstances beyond your control. This chapter will help you figure out what are the chances of success, given the current status of the organization. We assume here that you will do a splendid job in your role, rarely putting a foot wrong. We will focus instead on those things that may be beyond your control. These can make the task much easier—or make it "Mission Impossible."

This sizing up of the organization is not like an attempt to diagnose precisely a patient's illness. It is more like triage—the screening out of hopeless medical cases in order to avoid wasting scarce resources that could help others. The aim is to find out if the organization is beyond your help in the role you are considering. *Farmers don't plant their seeds in parking lots. People who seek change also should try to avoid wasting their efforts on situations where there is little hope of a fruitful outcome.*

This sizing up of the situation is often linked with a personal career decision—whether or not you should accept a new, challenging role within this organization. This makes it all the more important to do your homework and obtain a good understanding of the situation. You will personally earn the satisfaction of success or bear the brunt of failure, so it makes sense to avoid situations where many factors beyond your control are working against you.

The organization awakens to the need

Before major change can even be contemplated, top management must recognize the need and be sold on the idea. The dawning consciousness of a need for change may take various forms:

- The existence of a tantalizing opportunity that is just out of reach, unless radically new thinking is applied
- Signs of a decline, which everyone has some explanation for (from unfair competition to unreasonable customers or bad management), but which no one really understands
- An uneasy feeling on the part of the president that "we're not in Kansas any more, Toto."*

Usually none of these is sufficient. Some significant emotional event is normally required to concentrate the mind and crystallize the need for action. For example,

- The competition is eating our lunch.
- We cannot meet the payroll this month.
- Our most important customer just said "Good-bye."

This phenomenon—an inability or a reluctance to face new realities—occurs again and again. It is as old as time. Organizations usually do not heed the warning signs, do not grasp what is happening, and do not embrace change, until disaster stares them in the face.

This does not mean that no one understands what is going on. There are usually some people within the organization who see clearly what is happening and who are trying to create a change of direction. These people may become advocates for change and try to communicate their views to the decision makers—or they may become activists and try to create change within their own sphere of influence.

How likely are these people to succeed? There is no road map for the events leading up to a full realization of the need for change. This part of the journey is unpredictable, and it depends upon chance events and circumstances, such as

- External events, which may conspire to lull management into a false sense of security—or shock them out of complacency

---

* This is what Dorothy said to her dog, Toto, in the movie *The Wizard of Oz.*

- The skill and the positioning of people inside the organization who are advocating change, and the opportunities that these advocates can find or create to promote their cause
- The individual learning experiences of senior people, like who the president happens to sit next to at a gathering of his or her peers, or what business articles or television programs capture his or her attention.

The organization's journey cannot *begin until there is a recognition by top management of a* compelling *reason for change. This reason must outweigh the costs and the uncertainty of the changes required. Only when this occurs does major change become a possibility.*

It is usually around this time that management begins to look for help in figuring out what to do. Sooner or later someone will be asked to help make it happen. This person is a change agent, although top management (and the individual) may not at first realize this.

Avoiding a false start

One of the most appealing aspects of human nature is our endless optimism. As the saying goes, "Hope springs eternal in the human breast." However, optimism without information is often a recipe for disaster.

When you are contemplating getting involved in a change initiative, you have a moral obligation to assess the situation carefully and objectively, and to avoid encouraging the organization to attempt more than it can accomplish. This isn't only a matter of your career and your personal well-being, although that's clearly part of it. It is also a matter of avoiding harm to the organization and to other people's lives.

What is a false start? Picture a scenario in which management is trying to improve productivity by seeking to involve employees and by encouraging some of them to take part in quality improvement project teams.

> *After some natural caution and hesitation, a number of people get enthusiastic about the new approach, throw themselves into this style of working, and the projects begin to take off. The employees*

soon begin to smell success—they can see now the causes of many chronic problems that have been a source of constant aggravation in the past. They can see how to save the company a lot of money and reduce the daily hassle they have been enduring. They begin to feel a greater sense of self-confidence, a belief that they can actually nail these long-standing problems, and a new sense of pride in what they are accomplishing,

However, some senior managers begin to get uncomfortable with the situation. Although they too are anxious to see results, from their perspective, the situation is getting out of control. The teams are uncovering some major problems, and not just admitting their existence, but going around asking people for information about possible causes. The senior managers, who are used to evaluating and screening information before it goes outside their departments, see this behavior as publicizing embarrassing shortcomings. Also, the teams—through lack of experience—are overlooking some of the broader implications of their recommendations.

In addition, some department heads consider that their own people who are working on cross-functional teams are doing work for others. They also feel that their own departments are suffering from interference by team members from other departments. The department heads are concerned that these factors are undermining their ability to deliver on departmental commitments.

Unfortunately, these concerns have not been anticipated, and the operating style of the top management group discourages the members from raising such issues or trying to discuss them openly.

Some key senior managers, who have become ambivalent in their view of the teams, effectively withdraw their support. The teams begin to run up against unexpected roadblocks—team members are pulled back by their departments for "more important" tasks, and proposals by the teams are put on hold or receive a frosty reception.

The employees soon recognize what is going on and reluctantly accept that they can get nowhere in this situation. They are bitter at having their hopes raised only to be disappointed, and they consider that they have been deceived by management. There is an angry backlash and a loss of mutual trust and respect. The prospects of any positive change in the near future become remote, however desperate the need.

Setbacks and fumbles are normal and unavoidable; but, the organization that experiences a full-blown false start emerges worse off than if it had done nothing. Employees feel alienated and management's credibility is undermined. It is the change agent's responsibility to size up the situation carefully ahead of time, so this situation is avoided.

THE *MUST HAVES*

There are a number of ingredients that are essential to success in creating major change. Without these, the desired transformation usually cannot be started or, if started, cannot be sustained.

A compelling reason for change

Major change always entails unavoidable costs to the individuals involved: disruption, ambiguity, anxiety, and personal effort. In fact, these costs are such that major change will usually be resisted successfully and swept away unless there is a compelling reason for it to prevail. To those involved, the gain must outweigh the pain; the compelling reason for change must outweigh the costs of the transition.

Consider the following two examples:

1. "The president has staked his [or her] survival on the success of the quality initiative. We all are going to be under the gun to show our support."
2. "If we don't take action, the company will go under and we all will lose our jobs."

Which of these is the more compelling reason for change?

The first example is certainly reason to pay attention, but for many people this alone may not be an adequate reason to embrace change. The leader's visible commitment is vital, but it is not sufficient. *People throughout the organization have to understand the compelling reasons for change, and accept for themselves that change is necessary.* If people do not see good reasons for change, the leader's commitment may be discounted by many people as one person's pet project or eccentricity.

The second reason given—the possible failure of the organization—is indeed a compelling one, to which everyone can relate. If real change does not take place, everyone will suffer in some way.

Suitable sponsors

For major change to take place, there must be sponsors for the change—people who have the power to legitimize it. When the sponsors have granted their stamp of authority, the desired change is no longer a mere wish. It becomes company policy, and resources can be made available more easily. This does not ensure that the change will happen, but it is an essential first step.

What is a suitable sponsor? It is someone

- Who has the position, authority, and leadership qualities to be effective
- Whose style is consistent with the changes sought
- Who can articulate what change is needed and how it will be accomplished

The effectiveness of a sponsor is never just a matter of his or her position in the organizational structure. A position of authority is necessary, but a sponsor must also have the skills and personal attributes to be effective as a leader.

If this initiative were primarily a technical project, the style of the sponsor might not be such a big issue. However, when the change sought is itself partly a matter of management behaviors, then the sponsor's style matters greatly. *If the goal is to create a new approach to management, the necessary changes cannot be deployed using the old approach.*

Picture the following situation.

> *The president has read about Deming's ideas in an article, and something has struck a chord with him. He calls you excitedly into his office for a chat.*
>
> *"You know, this guy Deming is right about driving out fear. If our people had admitted to those problems in the northern region, instead of covering up for months, we could have taken action*

> *right at the start and saved ourselves millions. I'm going to make driving out fear one of my top priorities.*
>
> *"I know that you've tried to bring this subject to my attention before, so I would like you to take this on as your personal project. Please drop whatever else you are doing and focus on this for the next six months. My aim is to see a complete end to foolish cover-ups.*
>
> *"I know that it won't be an easy project, but you have my full personal support. I want you to keep me posted regularly, and if anyone is not giving you full cooperation I will deal with them immediately. This is a great opportunity for you to demonstrate the value of Deming's teachings. Can we get together tomorrow to discuss your ideas for going about this?"*
>
> *Now this president is a kindhearted person, but when he feels that others are less committed than he is to doing the right thing for the company, he cannot contain his frustration and anger. As a result, he is often very unfair to people who voice sincere and valid concerns about his chosen direction. He is known for chewing out people when they are simply trying to alert him to emerging problems.*
>
> *You leave the office wondering how to break the news to him that he has to start with his own behavior and the unintended signals he often sends by his actions.*

This particular sponsor is well-intentioned and is perhaps only a little out of tune with the type of change he is seeking. This person can be effective as a sponsor—if he recognizes the need for personal change and is willing to make the necessary effort. However, you would have little hope of success if, for example, the sponsor has threatening personal style or a profound mistrust of others.

Informed commitment of the sponsors

Let's suppose that the organization does have a compelling reason to change and that ideal sponsors exist. Obviously, all that is required now is to get the sponsors signed up for the change, right? Well, almost. The key point here is that *the sustained support of the sponsors will be required, not just a favorable decision now.* This means that the sponsors need to understand what they are getting into and be prepared for the long haul. They must not be misled by a rose-tinted picture of what lies ahead, or be encouraged to underestimate the magnitude of the task. If their commitment is not an informed one, the most likely outcome is a false start—withdrawal of the mandate at a later stage, and the inevitable collapse of the change process.

At this stage, even an informed commitment is still only an intellectual one. The people concerned have no practical experience of what they are committing themselves to. As the process unfolds, this intellectual commitment needs to be translated into personal involvement and an emotional commitment.

An adequate mandate for the Change Agent

If the management team does not recognize the need for a Change Agent, or does not understand this role, then the organization risks setting out on this journey with everything needed for success except someone who can read the map.

The Change Agent must obtain a mandate that recognizes that this role is critical to success. An *adequate mandate* usually means at least the following:

- Adequate time to devote to the job; that is, a full-time assignment or a limited set of other duties. (Chapter 7, "Internal Resources," explores in greater depth the issue of shared roles.)
- Full membership of the top management team, as a peer.
- An understanding that the change initiative will remain the responsibility of the top person—The Change Agent is simply there to help.
- A minimum time horizon of at least two years to get the process started.
- An expectation on the part of the top person that the Change Agent will become a close and trusted colleague. Although this is not written down, it is an important part of the mandate.
- A satisfactory formal definition of the role, including an understanding that the measures of performance for this individual are not short-term bottom-line.
- An adequate budget allocation.

THE CHANGE AGENT'S WISH LIST

The following ingredients create a starting situation that is a change agent's dream—one that includes more than just the basic necessities for success.

Top management credibility

In many organizations it is all too common for new initiatives to be announced and then allowed to wither and fail because of inadequate follow-through. When this happens, management credibility suffers and people learn to treat new pronouncements with a healthy dose of skepticism. When most people truly believe that top management means what it says, and will do what it says, then this credibility is a great strength and a bonus for the change process.

A cohesive top management team

In many organizations the existing management system rewards internal competition and, thus, undermines teamwork. If your organization's top management group truly functions as a team—supporting each other and working toward shared goals—then this is more than a bonus, it is a bonanza.

What does such teamwork look like? Perhaps top management members already do some of the following things:

- Agree on (and adjust) departmental budgets based upon the needs of the organization as a whole, rather than each manager trying to win the best deal for his or her own department.
- Set up cross-functional task forces or pool resources to tackle shared problems.
- Lend expertise and resources to each other to solve local problems.
- Treat each other with genuine respect and consideration.
- Give each other (and accept) moral support in difficult situations.

The informed commitment of the top management team

The informed commitment of individual key sponsors—above all, the president—is a must have. However, if the members of senior management function well as a team, they can act collectively as a sponsoring body, and this can be more effective and more enduring than the sponsorship of individuals.

If the top management team already understands the magnitude of the change process and realizes that it is its responsibility to make it happen, then it is *informed.* If the team understands the level of its involvement that will be required and is prepared to undertake this, then it is also *committed.* If the informed commitment of the top management team has been secured, one of the most problematic tasks is already done. As with the primary sponsor, this commitment starts off as an intellectual one, and the task remains of developing this into hands-on involvement and emotional buy-in.

Ready access to expertise

If your organization has ready access to relevant expertise (for example, instructors who can teach participative management skills or individuals experienced in process-improvement methods), then this is a great advantage. It means that you can tap into knowledge that you may not personally possess, probably without incurring exorbitant costs, and without fear of receiving a sales pitch from outsiders instead of objective guidance.

This type of expertise may be accessible from sister companies, from an enlightened corporate headquarters, or from a customer. You may find some of the experts you need simply by identifying people inside the organization who have been quietly working to improve things without a mandate from top management.

MIXED BLESSINGS

The following are situations that signal a need for caution and for careful investigation. Both represent possible threats and, at the same time, possible opportunities. They are not necessarily good news or bad news—you have to decide what you can make of them.

The organization is not autonomous

No organization completely controls its own destiny. All are subject to many external forces. However, an organization that does not have the autonomy to make its own decisions may not be able to carry through major changes, or may face greater risks in doing so.

Exhibit 2.1 gives a few examples of what can happen when strong outside influences are exercised in a destructive fashion.

Exhibit 2.1 *The risks of lack of autonomy.*

Headquarters (HQ) restructures
HQ restructures the organization and, in the process, removes the local general manager. Although the local managers still need and want to work together, their leader is gone and each now reports to a separate vice president in HQ. The team is torn apart by functional tensions.

Corporate staff functions feel threatened
In their efforts to improve, the managers of a local government department request changes to systems owned by corporate human resources and finance. These staff functions view such requests as challenges to their authority. The requests are refused and local managers find themselves viewed as troublemakers rather than trailblazers.

Board flexes muscles
A member of the hospital board becomes upset that so much time and money are being spent on staff (on providing training, holding events to recognize successful team efforts, and the like). The president can demonstrate large savings achieved by reducing error and waste, but is ordered to cut out such "extravagance," or lose her job.

However, outside influences need not be negative. Sometimes they can be very helpful. For example,

- Headquarters (HQ) may restructure or create pressures that support the change.
- The corporate staff functions may prove to be champions for change.
- The board may buy in strongly and help ensure the long-term continuance of the process.

It is essential to identify such outside influences and to figure out whether they are likely to be supportive, opposed to the changes, or neutral.

Some other major forced change is in progress

There are times when management has to focus all its attention on some issue that is critical, even life-threatening, for the organization, and almost everything else has to be put on the back burner for the moment. For example, perhaps external events force an immediate major downsizing. Perhaps a takeover or a merger is taking place. These may be times of crisis.

During such times, no one is going to pay the slightest attention to plaintive reminders about the improvement plan. However, great opportunities may exist, and some bold opportunism may pay off because

• People will grasp at anything that will help them accomplish the goal at hand. There is much less need for explanations, discussions, and cajoling.

"A formal process using planning tools might help us figure out how to execute the reorganization smoothly? Show us how!"

• Major decisions related to structure, strategy, and priorities are being made rapidly. Decisions that might normally require months of soul-searching may be dealt with quickly.

"Why not reorganize around key client groups or key operational processes? We never thought of that—Show us what you mean!"

"The merged organizations should find out what common strengths and weaknesses they have in their customers' eyes? Great idea, let's conduct some surveys as soon as possible!"

In such situations, the use of quality improvement principles and methods still can be introduced if this is done in the most effective way—as a means of achieving goals, rather than as an additional burdensome task.

POTENTIAL MAJOR BARRIERS

The following are situations that can present major barriers to change.

Massive current success

If the organization is currently enjoying great success, then complacency can become a formidable barrier—even if this success is clearly attributable to simple good luck or is the result of good decisions taken years ago. "We must be doing something right" becomes a kind of comforting mantra, used to ward off the unpleasant idea that fundamental changes might be necessary. In this situation, it can be very difficult to communicate the compelling reason for change, if indeed such a reason exists.

Full delegation of the process

The president may believe that his or her task is complete when the initiative has been approved and allocated resources. The words used may be *delegation* or even *empowerment.* The reality may be abandonment, caused by a lack of informed commitment by the principal sponsor.

A superficial or window-dressing motivation

Sometimes, what the president wants is not substantive change, but some activity that looks good to some key stakeholders, typically

Exhibit 2.2 *Assessing the situation.*

The must haves	Mixed blessings
• A compelling reason for change • A suitable sponsor • Informed commitment of the sponsor(s) • A clear, suitable mandate for the Change Agent You need all of these to succeed. Don't even think of compromising on these.	• The organization is not autonomous • Some other major forced change is in progress If either of these exist, find out whether they are mainly opportunities or threats, and consider whether they can be turned to your advantage.
A change agent's wish list	**Potential major barriers**
• Top management credibility • A cohesive top management team • Informed commitment of top management team • Ready access to expertise • Some good improvement initiatives are already under way If some of these exist, lucky you! Use these assets to the fullest. If they don't exist, your job is to develop them.	• Massive success • Full delegation of the process • A superficial or "window-dressing" motivation • Extreme dysfunction in the organization If any of these exist, check whether all of the *must haves* really exist.

shareholders or the board. The root cause here is usually not a lack of concern for the good of the organization, but a need for action to placate some stakeholder group, combined with a belief that this placatory action has no other value. In such a situation, this is probably all that can be achieved, regardless of how smart and hard-working the Change Agent may be.

Extreme dysfunction in the organization

Extreme dysfunction in the organization is the category in which we place all those internal problems that make it difficult to execute any major long-term project. Here is a sample: terrible relationships with employees and/or unions, rampant internal politics, functional fiefdoms waging open warfare with each other.

Any organization can be improved, but this type of organization is not ready to tackle the kind of major change we discuss in this book. If you choose to work with an organization that is not yet ready, you will first have to spend a lot of time on basic issues just to reach some level of relative stability and rationality. On the other hand, the very dysfunctionality of the organization may lead it more rapidly toward the kind of traumatic experience that will create an urgent desire to change.

To illustrate this, consider the "big three" U.S. automobile manufacturers. Each of these embraced a quality approach only when faced with the prospect of bankruptcy or humiliation in the marketplace—Ford in the early 1980s, Chrysler in the late 1980s, and GM in the early 1990s. One can argue that because Ford was initially in the worst shape of the three, it was the first to be forced to change and, as a result, it initially developed a significant lead over the other two. You may know of similar examples in government or in other sectors. So even extreme dysfunction may represent an opportunity—if you are prepared to wait long enough for the opportune moment!

HOW TO INVESTIGATE

In the rest of this chapter we will address ourselves to the aspiring change agent—someone who has been offered a job or someone who has no mandate but wants to create change. Other readers may find this material helpful to understand the process that a prospective job candidate may go through, and why.

When you are considering taking on the role of Change Agent and you know the organization and the people well, you already have much of the information you need to assess the current situation. If you are a newcomer or an outsider contemplating a job offer, you need to do much more information gathering. Regardless of your situation, here are some suggestions.

Write down the key information you need

You should write down everything you want to know, focusing on facts as much as attitudes and intentions; assign some priorities; and think about which sources are the best for each. You should also organize yourself to write down the answers you receive as you conduct your investigation.

Ask the simple, obvious questions

The simplest and most obvious questions are often the most revealing—and the easiest to overlook or assume the answers. Here are a few to put to your boss and colleagues.

- What do you want to accomplish with this change?
- Why do this now?
- What will be the consequences of not changing?
- What will success look like?
- What will you do personally in order to implement this change?
- What kinds of resources are you prepared to commit?
- What kinds of difficulties do you foresee?
- Who are strong proponents of this change, and why?
- Who are opposed to this change, and why?
- Why pick me as the change agent?
- What do you expect of me?

Use various sources of information

Here are some sources of information you can use.

- Your boss.
- Your colleagues.
- Employees at all levels. Frontline employees are a great source of information. (If your prospective colleagues are uncomfortable about you speaking to people at other levels, then this is an important insight into the organization's culture.)

- People who used to work for the organization, especially anyone who used to work for your boss.
- Key customers and suppliers.
- Key documents, such as the annual report and financial statements. *(Is the organization going down the tubes?)*
- The written description of your job. *(Has much thought been given to what type of person they are looking for? Does the job description seem to call for a miracle worker or a gopher?)*

It pays to be as thorough in your fact finding as time will allow. None of your prospective colleagues is trying to induce you to take the job by false pretenses, but equally everyone wants to ensure that you are aware of the attractive aspects. Also, they may simply not be aware that certain circumstances may be a problem in your eyes. For example, why should they disclose highly confidential information such as a pending takeover? They may not see why this could affect your job. But if the company taking over is known for its confrontational approach to employee relations, or an intensely short-term orientation, the takeover could stop the change process in its tracks.

Observe both words and actions

When you are already part of the organization, you have plenty of opportunity to observe people's actions. If you are an outsider, you may have to rely more upon impressions gained during interviews, but these also can be revealing if you listen and observe carefully. You can pick up many small clues, even in the course of just a few interviews. Exhibit 2.3 gives some examples of conscious and unconscious messages.

Things are often not what they seem, so you should not accept at face value everything you are told. How can you sort out the reality? Listen and observe carefully, trust your instincts, and probe further whenever you sense a disparity between the words and what your instinct is telling you. Usually there's nothing sinister going on; but, often people simply do not understand what the process is about, or the role you should play. You may need to reshape some of their expectations before you accept the job.

Exhibit 2.3 *Conscious and unconscious messages.*

Statement	Possible unconscious messages
I will do everything I can to support your efforts to improve quality.	This project is your responsibility, not mine.
In our division, quality is an integral part of everything we do.	We are already a quality outfit—We don't need to do anything different. I won't assign dedicated resources to working on quality, because it is already part of everyone's job.
This quality stuff is really just common sense.	People are making a lot out of nothing, with buzzwords and fancy-sounding methodologies. There's nothing here that we don't already know.
This quality stuff is really just good management.	I am a good manager, therefore I am already doing quality, even though I couldn't explain what that means.
Just tell us what to do and we'll do it —You're the expert.	Please don't expect us to think. We don't want any responsibility for decisions about how best to proceed. You can count on us to blame you for poor advice if things don't work in our departments.
If only we could communicate to our people how important quality is and get them to care about it.	The problem is our employees. We need a program to fix them. Management is doing a great job; besides, managers aren't responsible for quality—that's up to the people who handle the product and deliver the service.
This is a really exciting project/experiment.	I don't expect this to last very long.
I'm so sorry we had to reschedule this interview so many times.	Other things came up that are much more important than this.

Unconscious Messages

Be a skeptic

There's no need to become a cynic—to discard your faith in human nature or to suspect people of deliberately misleading you. But how many people are going to tell you that they are not in favor of improving quality, or that they consider the president's latest idea to be foolish?

Over-optimism, tact, self-preservation, maintaining appearances—these are all reasons why basically honest people may deceive themselves or become economical with the truth. So, it makes sense to start with the assumption that many of the must haves are missing. Then see if you can find any solid evidence to the contrary.

The president is completely committed? Great! But how much of his or her personal time is expected to be invested in making this work? Does the president realize that he or she will have to reexamine his or her own management style, just like all other managers, and learn some new skills? How does the president react to the idea of you coaching him or her?

Why does this management team want to undertake these changes? Is this reason sufficiently persuasive to justify disruption and extra work for at least a year or two? What will the personal consequences be for the management team if the changes do not succeed—a slap on the wrist, a slightly smaller annual raise, or a future on welfare?

Avoid self-deception

If the job being offered is really attractive to you—perhaps it offers a big promotion, a good raise, exciting prospects—then deep down you really don't want to discover that you are being offered a can of worms. It becomes easy to fool yourself, to overlook small but important clues, and to convince yourself that all is well. How can you avoid such self-delusion?

- Write down your understanding of what is involved in the job. Outline what you can deliver, what your boss and your colleagues are prepared to commit to, and what the expected outcomes are. Share and discuss this document. This exercise may help clarify the facts of the situation, the level of commitment, and your own thoughts on the subject.
- Talk the whole thing over with someone who has good judgment, who cares for your best interests, and on whom you can rely to tell you honestly what he or she thinks.

ARRIVING AT A DECISION

There is no mechanical way of taking all your observations and translating them into a decision about what to do—This is a matter of judgment. You can draw upon others to help you assess the situation and arrive at your decisions (see chapter 8, "External Resources"). You also can adjust your aims and expectations up or down according to what your starting situation looks like. If necessary, you can use the leverage you have (which is strongest *before* you sign up) to adjust your colleagues' expectations, or to negotiate conditions that provide a better chance of success. The important thing is to go into the situation with your eyes open.

WHEN A COMMITMENT TO CHANGE IS LACKING

If not us, who? If not now, when?

—Regarding change agents
Source unknown

We have been examining the situation as if top management had already made the decision to embrace change—hence the search for someone to help. Let's now look at the situation where no such decision has been made. It should be clear that without such a decision by top management, little solid progress can be made.

If you are convinced of the need for change, and you would like to do something about it, you have two principal options.

- You can curse top management's lack of vision, forget the whole thing, and perhaps look elsewhere for fulfillment in your work.
- You can try in various ways to move the organization toward readiness.

This latter option is *advocacy,* and it could take various forms, depending upon your position and role in the organization. For example,

- You could experiment with a quality approach on a small scale in your own department—in order to learn, to arouse interest and curiosity, and to build evidence that the approach works and produces results.

  Don't harbor any illusions—your efforts will be vulnerable, and the changes you make may not survive when you eventually move on, so this is not really implementation. But, your efforts can provide a showcase and, thus, be an effective form of advocacy.
- You may need to be patient and look for future opportunities. Perhaps you believe that the current situation is not sustainable and that a crisis will occur soon. This could provide the necessary incentive for management to contemplate changes.

  Perhaps changes in the top management team are on the horizon. If you can get some key senior people enthused about quality, they might make this the central plank of their strategy once they have the authority to do so.

• You can, in any case, work to recruit other advocates for this change, with the aim of influencing some key sponsors. Remember that advocates don't need to understand all the ins and outs of how the change will be accomplished. They only need to agree that change is necessary and believe that this approach is the right one. However, they must be able to influence the principal sponsors.

Building management commitment through advocacy

So, you decided to stick around and to work at building management commitment? Before we look at how you might go about this, let's deal with the possibility that you feel doubtful about your skills as an advocate. If you feel that you don't know how to do a great selling job, don't lose heart!

First, don't accept that you lack talent in this direction. Every human being can be persuasive, given something they really believe in. As the most successful salespeople demonstrate, you don't need to fit the stereotype of an extrovert with a massive ego. And you are not attempting a hit-and-run type of sale where bullying and psychological trickery are standard practice. Promoting a quality approach to your colleagues is more like proposing marriage than selling a timeshare.

Second, enlist help. If your idea is so great, some other capable people will also see the merits of it. Perhaps they are closer to the boss or have a better understanding of his or her thinking and priorities. Perhaps they will take the lead for you on this part of the journey.

You're worried that they might take all the credit? Well, now is the time to find out something important: *Do you really want to see this idea implemented, or is your first priority to be a hero?* This is one of the first personal decisions you need to confront when you become a champion for change. The issue of who gets the credit will come up again and again, so you should think about it now, and find out what your personal priorities are. You will not be successful unless you can quit worrying about getting credit for yourself, and focus on ensuring positive reinforcement for others.

Exhibit 2.4 *Building interest and commitment—management activities that can help.*

- Talk directly to customers about their views of the organization.
- Listen in on the customer complaint line.
- Find out how customers compare the organization with others; for example, competitors or similar operations.
- Estimate the opportunity costs of poor quality; for example, the likely impact of customer dissatisfaction on customer retention, reputation, marketing effectiveness, and market share.
- Talk directly to frontline employees about the difficulties they encounter in performing their work.
- Spend a day doing a frontline job—serving the customers or producing the product.
- Estimate the approximate internal costs of poor quality.
- Visit other organizations that are advanced in quality.
- Meet informally (for example, breakfast) with respected peers in other organizations that are advanced in quality.
- Study the application report of a quality award winner (for example, Baldrige)
- Study the criteria used to judge quality awards, and conduct an instant self-assessment.
- Take home selected quality-related videos to view in spare moments.
- Undertake a small, low-risk process improvement project on a management process, or take part in a simulation.
- Use some tools to map out a management process that cries out for improvement (perhaps the annual planning and budgeting cycle).
- Review the big picture and trends—existing business pressures, globalization, technology, and so on. Ask "What makes us think we will survive?"
- Ask "Can we survive using strategies focused only on *where* we are going—our goals? Don't we also need to have strategies for *how* we operate?"
- Estimate and quantify the likely consequences of doing nothing now. Develop a plan for dealing with the eventual consequences of doing nothing. This plan may be more costly and painful than a plan to begin change now.

Third, take time to develop, with your fellow advocates, some kind of plan for your campaign. It probably won't be precise and tidy, like a project plan with Milestone *B* completed next Tuesday, and it will evolve over time. However, it should be the result of some thought and discussion about

- The merits of the case and how these can be demonstrated
- Who are the most likely sponsors
- What are their goals and priorities
- How specifically your proposals will assist in the accomplishment of these goals
- What opportunities can be created to arouse sponsors' interest

At this stage, when there is no organizational commitment to change, you may not be able to propose actions whose stated purpose is education. However, you may be able to start promoting your ideas in a more subtle fashion. One of the most effective ways of educating people is to encourage them into situations where they will learn something from their experience. Perhaps you are trying to convince your target sponsors that customer relations are poor and deteriorating? Don't waste your time with rhetoric. Even hard data may not do the trick. Try to create learning situations, such as personal visits by senior managers to key customers, and see what happens. Experiences like this may create a new perspective. Remember that your colleagues want to do the right thing for the organization. It is their view of what is the right thing that you are seeking to change. Exhibit 2.4 offers some ideas for building interest and commitment among potential sponsors.

When you reach the stage of offering formal presentations or education of some sort, ensure that your ideas are always presented in a form that makes sense to your target audience and does not invite misunderstanding or negative reactions. For example,

- Avoid dogma. A sure way to undermine your credibility is to use blanket assertions that are hard to prove correct. You don't have to assert or prove that improving quality always improves productivity—you only need to demonstrate that there is a potential for productivity gains in this organization at this time.

- Minimize the use of terms, such as *total quality management,* that are jargon or buzzwords to most people. Although useful to practitioners, such terms are too easily misunderstood and misrepresented by others.

Using specialist terms with a nonspecialist audience also makes your message a hostage to the media. Remember that whatever is being promoted in the media today as the latest and greatest is likely to be the subject of derision within a year (or sooner) when the next popular fad has been created.

In chapter 8, "External Resources," we will discuss in more depth how to draw upon other resources to reinforce your message, and how to communicate the message in a way that is not vulnerable to distractions such as ill-informed articles in the popular media.

Being patient

It is unlikely that you can simply put together a neat presentation and sign up your key sponsors in one round. In fact, it may be foolish to attempt this because, if unsuccessful, it would close off further discussion for some time. Even worse, a premature attempt to close the sale might result in an uninformed decision, heading off in an entirely wrong direction.

You and your fellow advocates must be prepared to be patient and to spend as long as it takes—arousing curiosity, providing information to appropriate people in digestible amounts, and building a desire to examine the idea more thoroughly and formally. Sometimes this process takes years. Sometimes it seems a hopeless task, until some externally triggered crisis occurs.

But if your organization is not yet ready for change, you need not lose heart. *Every organization that is a leader in quality today was at one time not ready to begin this journey.* In every case, dedicated people within the organization acted as advocates, external events highlighted the need for change, the key sponsors were convinced, and the organization was ultimately successful in embracing new ways.

Pizza time

Whatever the outcome of your deliberations, there must be something worth celebrating by now.

- You decided to take the job? Congratulations! You have surely made the right decision, and you should celebrate the exciting opportunities that lie ahead.
- You decided to decline the job? Congratulations! You have surely made the right decision, and you should celebrate your narrow escape—avoiding something that wasn't right for you or that was an invitation to take part in a disaster.
- There is no formal change agent job on offer, but you decided to stick around and try to get the organization committed to change? Congratulations—you've got persistence! With patience and some luck, you might well succeed.

So, be happy with your decision, finish your pizza, and clean the sauce off your fingers before turning the page.

3

Getting Established

The secret of success is constancy to purpose.

—Benjamin Disraeli

CHAPTER CONTENTS

Getting established in the organization in the new role.

- Finding out how the organization works, and what's important
- How to investigate
- Building relationships with the leader, your colleagues, and other stakeholders
- Building a relationship with frontline people
- The importance of getting the cards on the table
- Agreeing on mutual roles, responsibilities, and expectations with others
- Agreeing on realistic goals—both tangible and behavioral
- Stepping up to the new role: personal education and development
- Working with people at a higher level than before

This chapter is addressed to the officially appointed Change Agent who must take the initiative in getting up to speed. However, it is in everyone's best interest that this individual become well established quickly, and this involves everyone's contribution. The leader, and everyone else on the team, should offer guidance, help, and encouragement as necessary. Now that you've accepted the job, you must commit yourself fully to the task ahead, for better or for worse.

In this chapter, we will deal with the tasks involved in getting yourself established. This is the process of progressing from being the new kid on the block to a valued member of the team—someone whose role is well understood, who has earned trust and respect, and who is appreciated as a contributor. Establishing yourself in this way is vital to success in your new role.

There is clearly more to this task than finding out where the washrooms and the coffee machine are located. If you are a newcomer to the organization, you will need to invest a considerable amount of time and energy for many months ahead.

Even if you have been with the organization for a long time, you still need to devote time to this kind of groundwork because this is a new role for you. The people in the team may be familiar to you, but, there are new expectations to be agreed upon and different working relationships to be forged.

Because you sized up the situation carefully before accepting the job, you already have some good information about what is going on and a list of issues to explore in more detail. At this stage, your main objectives as the newcomer are

- To clarify the organization's goals
- To figure out how you can make the greatest contribution to these goals
- To clarify your role and to set mutual expectations with your colleagues
- To start building working relationships with your colleagues and other stakeholders, and to gain a deeper understanding of their personal outlooks and aims
- To reaffirm and work out some of the tangible commitments that are prerequisites for success (for example, senior management accountabilities and budgets)

Getting established takes time, so it cannot take place as a separate exercise, like a preliminary to starting work. It is best carried out in the context of doing the work. You have to establish yourself in the process of working with your colleagues to plan the transformation.

We will look at the process of getting established from three points of view.

- Understanding how the organization works
- Building relationships
- Clarifying roles, responsibilities, and expectations

UNDERSTANDING HOW THE ORGANIZATION WORKS

If you are a newcomer to the organization, you have a great deal of work to do in order to get up to speed, and learning how the organization works is part of this. If you are a long-serving employee, you should still review your level of knowledge and understanding. This section outlines some of the things you need to know.

What you need to find out

Here are some key areas you need to understand.

- What is the formal structure of the organization?
- What jargon is used (acronyms and special names for products and services, features, methods, technology, standards, operating procedures, customer groups, special committees, and so on)?
- What is important to success in this type of organization? In transportation, it might be the scheduling of aircraft or trucks. In a producer of high-technology products, it might be time-to-market. In a health care facility it might be reduction of the patients' length of stay.

- What are the chronic issues and problems that just never seem to go away, that are viewed as facts of life in this type of organization? Some of these are "low-hanging fruit" —easy to harvest once a rational, organized, and process-oriented approach is adopted.
- How does the organization function as a system, and what are the key processes (for example, the activities that consume resources and/or are critical to success)? If you can draw up some kind of simple, high-level process diagram you are well on the way to understanding this dimension. Don't forget customers, distributors, suppliers, and so on.
- How does the organization see itself? What are its history, traditions, and mythology (stories of the deeds of the founders, or whatever)?
- What are the unwritten rules of behavior (for example, "Always keep subordinates in their places—A senior person needs to demand respect.")?
- What unwritten rules for survival exist? For example, "Never admit you're going to miss a project milestone" (someone else in the project may be forced to admit this first), or "Always spend all your budget" (or you will get less next time around).
- What are the internal politics? You are not going to play politics, because this would undermine your ability to win the trust of others; but, you need to be sufficiently aware of what's going on around you. You need to know what political threats exist and what hidden agendas may disrupt the agenda for improvement. You risk appearing naive, and therefore less credible, if you are unaware of this dimension of what's going on.

Identifying the key processes

Understanding the organization from a system or process point of view is important and very helpful. Here are some ideas for gaining this understanding.

Virtually every organization has a set of *management processes.* Regardless of the type of company or institution, these processes fall into a common set of categories. These categories may be thought of as subsystems within the overall management system. Exhibit 3.1 (adapted from the Baldrige criteria) illustrates this "universal anatomy" of management systems.

The processes used and the degree of importance attached to each subsystem—for example, the degree of focus on customers' needs—may vary from one organization to another. However, most of these subsystems exist in some form in almost every organization.

Every organization also has *operating processes*—those activities that are directly responsible for delivering goods and services to customers. These operating processes differ greatly from one type of organization to the next. Exhibits 3.2 and 3.3 are sketches of some of the key operating processes in two types of organization.

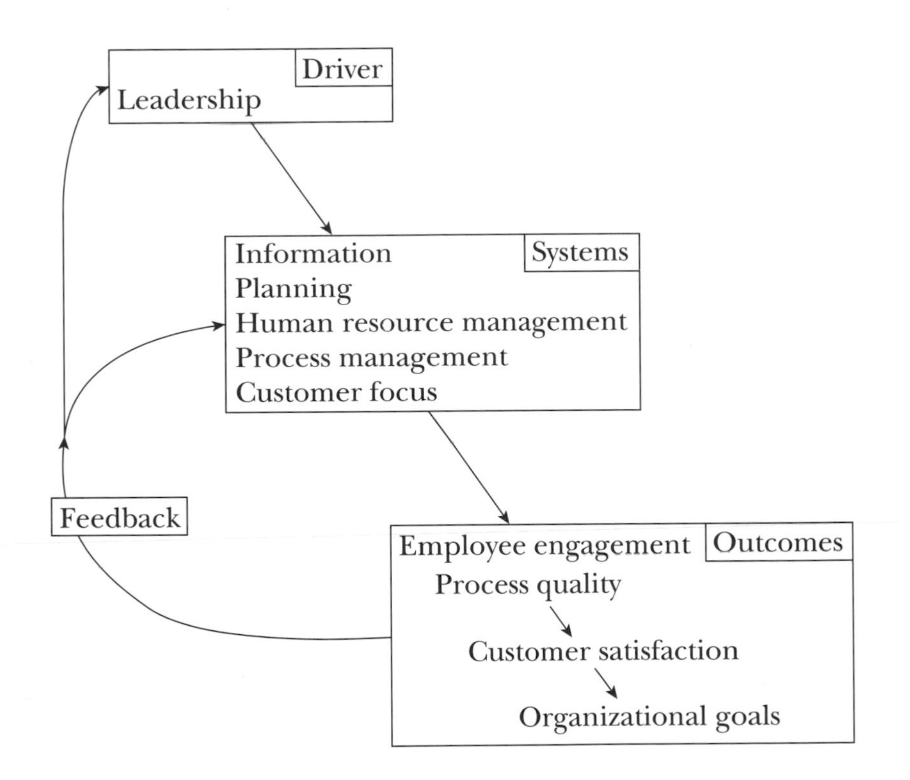

Exhibit 3.1 *The anatomy of management systems.*

AIRLINE

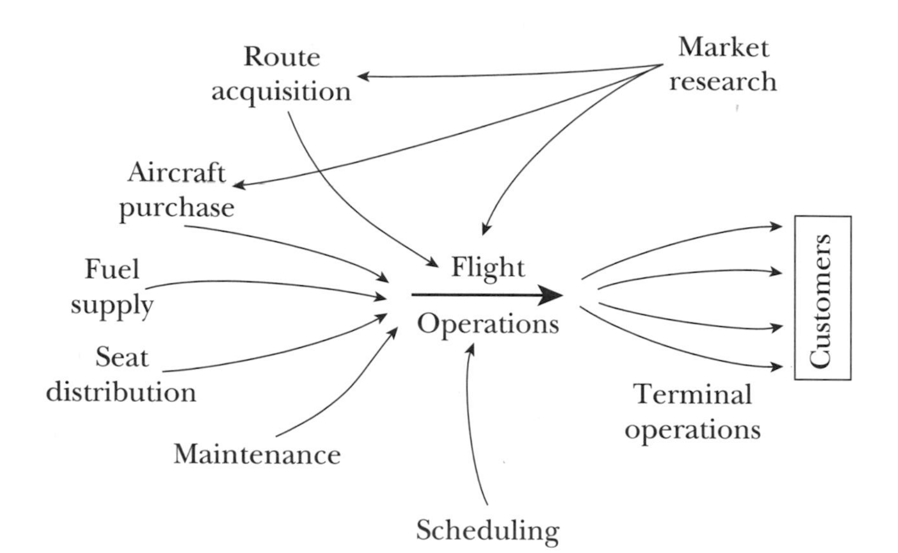

Exhibit 3.2 *Typical operating processes for an airline.*

AUTO MANUFACTURER

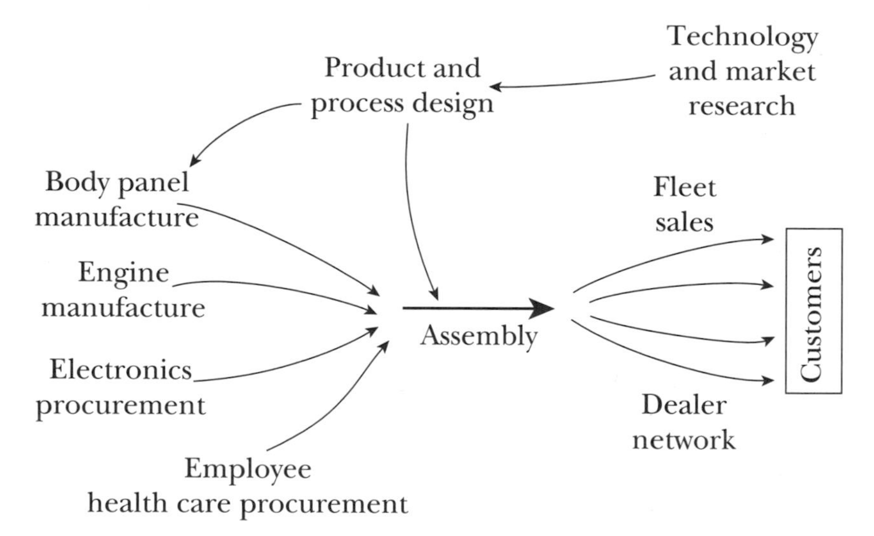

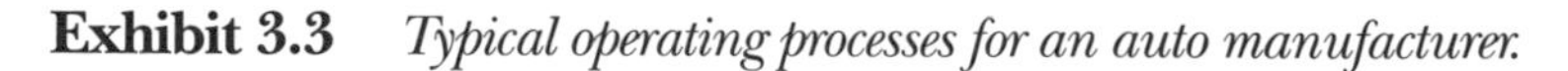

Exhibit 3.3 *Typical operating processes for an auto manufacturer.*

The management processes may be thought of as *support* processes. They are necessary to support the operating processes, which are the core value-adding activities of the organization.

How to investigate

The ideas for assessing the situation set out in the previous chapter are still relevant here, but now you can escalate your pursuit of information and understanding. This should be fun.

For most types of information, there's no better method than just asking directly. Asking pertinent questions is one of your first contributions to the organization, because this provokes thought. Even the simple question "Why do we do this?" is often hard to answer. To be successful in your pursuit of understanding *you should never be afraid of looking stupid, and you should never feel apologetic for asking or for not knowing already.*

While carrying out this investigation, also remember that *in this job you are always scouting for volunteers.* You are always looking for people to sign up for the cause of making things better. The people who are most candid about what's wrong are often strong candidates. If they want to make a contribution—and most do—then in time it should be possible to find roles for all of them.

What success looks like

You are successful in understanding the organization when you can

- Talk to people within the organization in their own language
- Demonstrate a grasp of the important issues that concern people
- Understand the organization and its component parts as a system, and thus have a feel for the underlying dynamics and the possible levers for improving quality and productivity

When you reach this stage, you may find that you can offer some insights or a useful new perspective for your colleagues. At the very least, you should now command some respect as someone who has done his or her homework properly and who knows the score.

BUILDING RELATIONSHIPS

Your leader

One of your first priorities—now and in the future—is to sustain the president's sponsorship of the change process. In time, it should be possible to secure the collective sponsorship of the entire management team, but for a while everything will hinge upon the leader's personal resolve.

Your relationship with the leader is somewhat different from the relationship with your colleagues because the leader's role is different. This individual faces different pressures and has different priorities from anyone else. This is the one person who identifies only with the success of the organization as a whole. Also, because the president is the top member of the company, he or she should have no need or desire to play politics within the organization.

In other ways your dealings with the leader will be the same as with other colleagues.

• You need to understand his or her priorities and preferences, the external forces that impinge upon him or her, and how you can help this individual achieve his or her goals.

• You need to establish a relationship of mutual trust and respect, just as with your other colleagues. You need to maintain the leader's full confidence in your integrity and competence and his or her full support for the process.

However, you should not seek any kind of privileged insider position that gives you some perceived advantage over your colleagues. Favoritism is incompatible with the open team approach you will be pursuing. If it is clear to everyone that you, like they, are judged solely upon the merits of your contribution, you will be in a better position to secure the trust of other colleagues, to discover their goals and concerns, and to act as a resource to help them.

• You need to help the president understand *in practice* the transformation he or she is leading the organization into.

You need to provide the president with opportunities for learning that suit his or her preferences. You should protect the president from needless surprises and embarrassments as he or she too struggles as a novice with new ways of working. No one likes fumbling and stumbling; but when you are the head of the organization, you may feel that you are letting your people down when you appear foolish.

Finally, as you build your relationship with your leader, you need to remember the isolation and the immense pressures of this person's position. If you feel that your job is tough, you should have some empathy for the person who sits in the office where the buck stops.

Your colleagues

Your aims, at this stage, in building relationships with your colleagues are probably

- To gain an understanding of their outlooks and personal goals
- To explore how you can help them achieve their goals
- To begin to get some sense of who may emerge as the strong champions for the process, and the concerns that may get in the way of those people and others coming on board

Although you may have informal or purely social get-togethers with your colleagues, relationships are mostly built in the course of working together day by day. So, while you are exploring and negotiating agreements with your colleagues about roles, responsibilities, and expectations, you are also setting the tone for your future relationships with them.

Building a relationship takes time and effort, and it starts with a genuine desire to understand the other person's point of view. For the Change Agent, this desire should be genuine because understanding the other person's point of view is a prerequisite for success in making the process work. *The only reason for any of your colleagues to support the change process is if this helps them in some way to achieve their goals.* You must therefore find out as best you can what these goals are, so that the change process can be used as a means of achieving these.

Here is a model of how a relationship may develop over time.

- You each share your work-related objectives, responsibilities, and problems.
- You begin to share your personal values, ambitions, goals, and concerns.
- You identify common ground and build on this—from related work goals to shared interests, likes, and dislikes.
- You begin to look for ways in which you can help each other achieve both business and personal goals.
- You begin to establish a pattern of regular contact that is productive (and preferably pleasant).

There is some key information about each of your colleagues that you need to establish as soon as possible.

- What is there about the proposed changes that they find attractive? What may benefit them?
- What is there about the proposed changes that they object to? What threatens them?

Some of this may be sensitive information that the individual concerned will only share with you when he or she is convinced that his or her confidence will be respected. For example, a manager may find the prospect of empowering employees very threatening, or may be very attached to the symbols of prestige and power.

As you get to know your colleagues, and a relationship of mutual trust and respect is built, the conversation may sometimes turn to a discussion about the other members of the team. When this happens, you are on dangerous ground. Discussions like these are not in the spirit of an open, team-based style of working. In fact, they run the risk of developing a conspiratorial or disrespectful atmosphere. You should steer away from this type of discussion.

However, people often feel a real need to share their feelings and perceptions about their peers—especially when relationships are strained—and to get feedback from others on how they are perceived. Rather than taking part in private discussions, a better approach is for you to encourage the use of practices that enable everyone (including you) to share more about themselves and to learn how others perceive their contribution. Exhibit 3.4 offers some ideas on how to do this.

During your discussion, you will begin to discover some of the concerns people have about the proposed changes. As your colleagues gradually open up and share some of these concerns, you need to acknowledge the validity of their opinions and find out what lies behind them.

For example, if someone says "This quality improvement program resembles things we've done before that were just feel-good exercises," you must not instantly leap to the defense of "your" program, or interpret this as an attack. You need, first of all, to find out what the person really means, and why he or she feels like this. To illustrate, here are just two of many possible interpretations of the previous statement.

1. This person fully understands the magnitude of the changes required to revitalize the organization and wants to see these changes happen, but fears that the president only wants a superficial, cosmetic exercise. He or she wants to see some substance this time.
2. This person believes, based upon experience and the information he or she currently has, that quality improvement inherently lacks substance and doesn't amount to much more than a feel-good exercise.

Only when you understand the basis of the other person's views can you begin to establish common ground and offer information that may win them over.

These exploratory discussions with colleagues should not focus only on the change process. They should start with the larger picture, from the other person's perspective, and work down to individual issues and your perspective.

You may feel that the change process is your world—the reason for your existence in the organization. To your colleagues, however, it is just one issue among many. You need to link the change process to your colleagues' larger concerns, so that it becomes a means for them to achieve their goals.

Exhibit 3.4 *Tools for building relationships within the team.*

One of the most effective ways of starting to earn the trust of your colleagues is to be very open in sharing who you are—and being very open to feedback from them about your actions.

You can use tools to do this that can accelerate your acceptance by the team and encourage the use of methods that will help bind the team together. For example,

- You may use personal self-assessment instruments, like FIRO B or Myers Briggs, to quantify the "shape" of your own personality. (If you are not familiar with these instruments, your human resources colleagues should be able to help you.) These instruments provide insights into our styles of thinking and working, and our preferences and biases. Sharing your own results with colleagues can be a good way of building mutual understanding—and they may in turn share information about themselves that will help you learn to work together.
- You may use tools to get feedback from several directions—from your leader, peers (who are also customers), and people who report to you. This is sometimes called *360 degree feedback.* By using such methods, you can open up valuable channels of communication with others, demonstrate your openness to feedback, and set an example that others may decide to emulate.

Peer feedback and upward feedback are particularly valuable in helping people to understand better how they are perceived and how their behavior affects others in the workplace. If your personal example encourages others to adopt such an approach, then you have already accomplished one of your objectives (establishing a behavioral feedback mechanism) without making a ""big deal" of it or creating a situation that feels threatening to your colleagues.

This broader focus should continue to apply in your future dealings with your colleagues. If you focus on the change process too much, you run the risk of becoming isolated and disconnected from the running of the business. You must avoid becoming focused on the change process for its own sake—it is only a means to an end.

Here is a test of your judgment in this respect. Can you answer the following question: Under what circumstances would you completely drop all work on the change process in order to help your colleagues contain some crisis?

You *should* be able to name some possible circumstances, because sometimes you must drop everything else to help in a crisis. As a full member of the top management team you also share the team's broad responsibility for the health and survival of the organization.

If the boat is in imminent danger of sinking, you must help your colleagues operate the pumps, even if this means temporarily abandoning your post as navigator. *Or, would you rather take professional pride in the fact that the vessel sank while exactly on course?* There is little point in devising the world's greatest improvement system if it doesn't help the organization to be more successful, or if the organization goes out of existence before the system can be implemented.

Getting the cards on the table

Picture yourself as the newly appointed "Vice President, Continuous Improvement" for International Services. Here is the scenario:

> *When you report on your first day, you find that your boss, the president, has called a special meeting of the top management team. During the meeting he introduces you in a very complimentary fashion and expresses his full confidence in your abilities.*
>
> *He explains your role, and sets out his commitment to quality in a passionate and eloquent manner. He stresses his determination that the organization should go in this new direction, leaving no doubt in anyone's mind of the strength of his conviction. Finally, he throws the floor open for comment and reactions.*

> *Your new colleagues, two of whom you met during the interview process, welcome you courteously. Most also take the opportunity to summarize their position regarding the new direction. It soon becomes clear that not everyone has been infected with the president's enthusiasm.*
>
> *While no one overtly opposes the new direction, several individuals openly admit that they have some concerns about it, or that they really don't know enough to be able to form a judgment. This seems to come as no surprise to anyone—these feelings obviously have been voiced to the group before. Others offer you their support along with the hope that this "experiment" will be a great success. One person is obviously firmly opposed to the whole idea, and does little to conceal his feelings.*

How should you feel about this? You should be very pleased, because you have a group where people are prepared to declare where they stand. The cards are on the table. This is a rare and highly desirable starting situation. Often it takes months or years to achieve such openness, and sometimes it is never achieved.

"Open opposition is good?" you ask. "I'm supposed to feel that I've died and gone to heaven because people are throwing brickbats at me?" In a word—yes. The key point is that *it is highly desirable to achieve the open sharing of concerns.*

Your colleagues on the team are not experts on quality—how could they be? That is why you were hired. They may not really understand yet what they are getting into, and they almost certainly do not know enough yet to make an informed commitment. Every single one of them—even the president—has some doubts and concerns. It would be extraordinary if they did not. The last thing you want at this stage is support that can only be a facade. Lack of team discipline or inability to build a consensus are weaknesses. However, openness, as displayed in this scenario, demonstrates honesty and mutual respect, and provides the opportunity to open a discussion on the issues that your colleagues care about.

In reality, it is much more common for concerns to be concealed—because people do not want to expose their doubts and lack of knowledge, or because they do not want to declare their concerns openly for fear of appearing negative. Hidden concerns are like hidden agendas—difficult to deal with and likely to undermine the functioning of the group.

You should strive to get your colleagues to open up—perhaps one-on-one at first, and then as a group—so that the issues that really concern them can be thrashed out. You will be more successful in doing this if you

- Win their trust by demonstrating integrity and by being prepared to make yourself vulnerable
- Respect their concerns as legitimate and sincere, and respond accordingly
- Offer information and ideas, rather than dogma
- Treat each person's situation and perspective as unique, even when their concern feels like a standard one to you
- Don't feel obliged to have all the answers or to be right all the time
- Welcome open debate as a useful part of your ongoing relationship

You also need to be a careful listener, because people ask for information in different ways. For example, someone might say to you, "I just don't understand what improving quality will do to our costs." Or, someone may throw out an apparent challenge like, "We cannot afford better quality. Our costs are already too high." This may simply be this individual's way of broaching the subject. Another individual may offer statements of support that conceal their own doubts or misunderstandings, like "I am sure that the extra costs of improving quality will be more than justified."

The last two statements both suggest an underlying misunderstanding. Both are invitations to open a discussion and provide information, but you need to be listening carefully and nondefensively.

Employees

The idea of developing relationships with a few colleagues is easy to grasp. However, the organization (or your part of it) may have dozens, hundreds, or even thousands of employees. If you have a senior position, you may be asking yourself, "How can I have any kind of relationship with our frontline employees?"

The answer is that you inevitably will have some sort of relationship with these people as a group, and hopefully relationships with many of them as individuals. These relationships may be shaped by remoteness, fear, and suspicion, or by contact, empathy, and trust. And these relationships are vital. *You and your colleagues have little chance of success in improving the organization unless you earn the trust of employees and persuade them to apply their talents and energies to shared goals.*

As soon as you are appointed, employees will begin to form a view of you as a person, and of what you are trying to accomplish. Whether or not you take steps to communicate, these views will soon solidify and are then hard to change. If you don't communicate with employees, you are still sending a signal—that they are not important enough to merit your attention.

Who are you to employees? You are an ambassador for a new style of management that the organization needs and that most frontline employees will welcome. You are someone who realizes the impossibility of doing a great job when the tools and information are inadequate, and when no one will listen to employees' job-related concerns. You are someone who shows respect to others regardless of rank, and who will not indulge in deceit and manipulation. To them you are someone who understands their perspective and who can be trusted.

So, you must set out from day one to establish this relationship—one of empathy and trust—with employees and set an example that your colleagues may emulate. How can this be done? You face a few obstacles:

- The sheer numbers may make it impossible for you to get to know more than a fraction of employees as individuals.
- You need to avoid undermining your colleagues, thereby losing their trust and your ability to influence them.
- It will take time to bring about significant changes in the management system. In the meantime, many important employee concerns cannot be addressed effectively.

You can succeed by

- Spending time with employees on a regular basis—and listening more than talking.
- Being open about who you are—your attitudes and beliefs.
- Listening to concerns, trying to understand, and showing empathy—even though you often cannot fix the problems immediately.

- Looking for ways to signal trust in employees and concern for their well-being. Actions like scrapping "dumb" rules or sharing previously secret information can have great symbolic importance, and may cost nothing. Even small gestures of consideration can earn goodwill out of all proportion to the effort required, as illustrated by Exhibit 3.5.

Another aspect of the relationship between you (and your colleagues) and employees is the nature of the first formal actions taken as part of the improvement efforts. You should avoid, at all costs, actions that signal to employees that *they*—their attitudes, their lack of care and attention, or their lack of motivation—are the problem.

For example, poor telephone answering is often identified as an irritant to customers, and may be chosen as an "easy" area on

Exhibit 3.5 *Small gestures signal big changes of attitude.*

"One of the best things I ever did after I joined the organization as the Change Agent was seemingly trivial. I was talking with an employee in his work area when I began to notice a buzzing in my ears. When I commented on this, he explained that there was an air vent nearby that made this strange noise. 'Doesn't this drive you nuts?' I asked. 'It sure does,' he replied, 'but it has been like that for years, and the maintenance people are busy working on more important problems.'

"On an impulse I called up maintenance and asked them very politely whether they could fit this small task into their schedule. For me, they were most helpful. The job was done within a half hour, and the employee was delighted.

"I was surprised and pleased at this gratitude, but the implications went much further. This individual was from then on one of my most fervent supporters.

"The fund of goodwill I earned by small gestures like these never seemed to be used up. This helped a great deal in situations where we were asking employees to accept unpopular changes, or when we made mistakes and had to mend fences."

which to start work. The usual approach to this issue is to establish some performance standards (for example, every call to be answered within three rings), and then to expect these standards to be met *without any method being offered for achieving this.* There is often a confident assumption made that this is just an "attitude problem." By telling people about the standards, and by establishing and publicizing measurements of performance, employees will be encouraged to "clean up their act" and just pick up those phones when they ring. This arrogant assumption seems to be confirmed when most of the top people achieve almost perfect scores, while others struggle, and some departments do not seem to improve at all.

The reality is that there are countless reasons why phones may not be answered quickly, such as people being out of the office on business, overload of the telephone lines or the people, work that demands frequently leaving the workstation for a few moments, work that requires stretches of uninterrupted concentration, lack of training or information on how to use features like call forwarding and call pickup and so on. Most managers are affected by some or all of these restraints. They would laugh at the very idea of attempting to answer their own phone every time, every day, within three rings. Yet, they may simply pass the problem onto their personal secretaries and wonder why everyone else in the company cannot meet these standards.

This approach illustrates a perfect way of unintentionally alienating employees—and killing enthusiasm for quality improvement—by signaling that employees are the problem. There's nothing wrong with tackling such an issue—if management adopts an approach that looks more like this.

- Accept that the problem may be more complex than it appears at first glance.
- Accept that most people are already doing the best they can—the bulk of the problem is unlikely to be caused simply by a poor attitude among employees.
- Set out to discover the barriers that prevent employees from achieving the performance goals.
- Offer some methods—such as the creation of teams to tackle the issues, training in problem-solving tools and techniques, and facilitators to coach and support these early efforts.

This approach makes sense, helps people, and will work. It will be welcomed by employees, and as they learn to fix the problems they will be proud of their accomplishments.

Building relationships with employees calls for more than being a good listener and a sympathetic person. It calls for an understanding of how to fix the problems within the system, rather than unconsciously blaming employees and setting out to "fix" them.

Other stakeholders

There are usually many stakeholders who can have a major influence on the success of the change process. You need to identify the key players and decide how to establish contact with them. These stakeholders may include

- The board, or some key members thereof. Naturally, any contacts you make with members of the board will be with the knowledge and agreement of the president, who has to manage relationships with this group of people.
- Union representatives.
- The people who championed previous and existing improvement initiatives.

Let's consider some of these stakeholders in more detail.

Union representatives

When there is an established union in your workplace, there are only two possible ways of implementing a quality approach: to do so in partnership with the union representatives, or to break the power of the union in your workplace. There is no middle ground.

We will discuss only the partnership route, because taking the other route is a dangerous strategy, usually born of desperation, and likely to fail. The following are some ideas regarding how to make union people partners in the process. This may seem an unlikely prospect, given the current relationship, but it has been done successfully by many others, with unions no less tough-minded than yours.

- *Understand and respect the union agenda.* This is usually related to issues such as job security, health and safety, and the work environment—issues that are important to your people and that you should also care about.

• *Look for common ground.* You will likely find this common ground in issues to do with quality of products and services and the work environment. The area of job security is also common ground in one sense, because no one wants the company to go out of business. But job security is usually a contentious area, because the union people feel compelled to take an incremental view and fight for each individual job. Management, on the other hand, has to take a macro view, and strive for higher levels of productivity that enable the entire operation to be competitive.

• *Forge some regular communication vehicles.* Your aim here is frequent open sharing of relevant factual information and regular personal contacts. These will enable you to build mutual trust and an honest relationship centered on the areas of common ground. Typical communication vehicles are a regular newsletter or fact sheet, and a regular management-union meeting.

The union-management meeting may be explosive at first, and this won't change overnight; but if the management team is serious in their intent, it must persist. Only through regular personal contact can people get to know each other well enough to establish mutual trust. It is unrealistic to expect the honeymoon to start immediately—there is a legacy of mutual antagonism and mistrust that reaches back far into history. The unions have excellent reasons for mistrusting management and for automatically viewing management overtures toward them as devious tactics.

• *Get close to your frontline people.* Union representatives are elected and the frontline employees are the union's power base. If management can change its own relationship with frontline people, this will in turn change management's relationship with the union.

The aim is not to circumvent the union people, which would pose a threat to them. The aim is to ensure direct personal contact with frontline people and to build a relationship of mutual trust and respect. Management needs this relationship in order to tap the human potential that often lies dormant.

• *Deal with facts and confront reality.* By seeking data and focusing on the facts in every situation, you can move away from emotional win-lose confrontations, toward compromises that respect everyone's needs and take account of reality. Sharing and agreeing on the facts is also a big step toward establishing mutual trust. It helps each individual to understand the other person's viewpoint and be able to put themselves in others' shoes.

The reality is that you are in this together, there is common ground, and you can help each other by cooperating. The enemy is not each other, but costly waste of human effort and materials, and poor work that drives the customer away and thus loses jobs.

- *Maintain your sense of humor and try to make it fun.* In order to build a relationship with people who currently view management as the enemy, you will have to make yourself vulnerable at times and be prepared to take some knocks. A sense of humor can help you to keep things in perspective, defuse attacks that begin to feel personal, bring overheated conversations back down to earth, and lighten up the tone of meetings. You will find that the people on the other side of the table have a sense of humor, too.

Champions of previous improvement efforts

It's easy to overlook these people. Often, past efforts are perceived to have failed in some way, and those still under way may be perceived to have shortcomings. Perhaps this is part of the reason why a more formal approach is now being adopted. *Don't attribute the limitations of past improvement efforts to shortcomings of the people involved, and don't underestimate the importance of these people to future efforts.*

It is likely that the people who took the lead in these previous initiatives did the best that could possibly be done in the circumstances. Did you ever try to run a quality circle while supervisors were trying to pump up the volume regardless of error rates? How do you rate your chances of improving customer satisfaction and loyalty when the salespeople are being measured and rewarded on nothing but monthly quotas?

The people who supported such previous efforts—and perhaps failed, or were not recognized for their successes—are the very people who have the courage, the motivation, and perhaps even the knowledge and expertise to make this latest initiative successful. Some of them may have succeeded in circumstances much more difficult than you now face.

These people would probably love to have another chance at creating change, with the sheer luxury of management support, a proper mandate, and proper tools and training. If they are involved in initiatives that are currently under way, you may hold the keys to making these truly successful.

The worst thing you can possibly do is to "wipe the slate clean," forget the past, and launch a brand new initiative as if history had not happened. Instead, build on past efforts by including the people who were involved.

Go out and find these people and begin to sign them up, if only as moral supporters at first. You may want to recruit some of them as soon as possible onto your own team—the support network.

Your team

You will develop your own team—a network to support the change process—and you may also have people reporting directly to you. You need to build your relationship with these people, just as with your other colleagues. The difference is that it is your responsibility to set the tone and to focus attention on what is important. Your initial approach may look like this.

- Spend time with these people, one-on-one.
- Share your personal vision (for the organization and for your team) as a starting point, and get your team's input as to what it should be.
- Share your personal values and who you are (you may use the tools we mentioned).
- Agree on operating principles or values for your conduct as a team.
- Set out to "walk the talk," and set up regular feedback to ensure that people (including you) find out when they are straying.
- Focus attention on the organization's goals, and share ideas on how best to support these.
- Involve your team in efforts to explore your internal customers' needs and wants.
- Involve your team in developing a plan that supports the management team's plans and, hence, the goals of the organization.

Picture your relationship with these people as a model for the relationship you would like to have with your leader. If you want regular one-on-one meetings with the president, why would you deny your people this consideration with you? If you feel it will help the president to have a feedback mechanism and coaching on his or her behavior, why not arrange for feedback from your people in the same way? You may be more sympathetic to your leader's struggles and frustrations if you are going through similar experiences with your team.

Remember that you must be a role model, and this almost certainly involves significant changes in your outlook and behavior, too. Perhaps you feel that you are already a role model. Perhaps you are not aware of any way in which your style of working needs to be improved. If so, perhaps you are not the right person for this job.

Exhibit 3.6 offers some guidelines for building relationships with colleagues—whether these are senior managers, frontline employees or union representatives.

Exhibit 3.6 *Guidelines for building relationships with colleagues.*

- Stay focused on the common goals.
- Strive for the highest standards in your work for others.
- Demonstrate a generosity of spirit: Be tolerant and always give credit for accomplishments.
- Be positive and enthusiastic.
- Always assume positive/honorable intentions until proven otherwise. Better to be disappointed sometimes than to undermine all your relationships with negative assumptions.
- Always show respect for others and trust them to do what seems the right thing in their eyes.
- Be open. Share with others who you are and what you stand for. Share concerns rather than bottle them up.
- Maintain your integrity and demonstrate that you can be trusted—this is one of your most valuable (and perishable) assets.

AGREEING ON ROLES, RESPONSIBILITIES, AND EXPECTATIONS

In this section we will discuss the process of reaching agreements with your leader and your colleagues that define what you are collectively trying to achieve. This section also addresses who will do what in order to make this happen.

Getting established also has an important dimension that has to do with relationships: building the mutual understanding, trust, and respect that are necessary for you to work effectively with your colleagues.

We will discuss these two dimensions separately, but they are closely intertwined. For example, any misunderstandings regarding commitments are likely to spill over into strained relationships and mistrust. On the other hand, lack of mutual trust can render even formal agreements worthless.

This task is not complete after the first pass. Your role will continue to evolve over time as the needs of the organization evolve, and as understanding of the change process grows. You will often need to return to the discussion of how you can best contribute, and keep adjusting your priorities in order to satisfy the changing needs of your (internal) customers. That is the nature of the job. Exhibit 3.7 suggests one approach to clarifying mutual expectations with your colleagues.

Typical questions regarding the situation

Here are some of the issues that you began to explore before taking the job, and that you will now want to clarify further with your colleagues.

- What does the organization hope to accomplish? In what time frame?
- What do they want you to accomplish? By when?
- How is success defined? What does it look like?
- What budget and other resources are allocated or envisioned to support the change process?

Exhibit 3.7 *Clarifying mutual expectations and obtaining feedback.*

An effective way for the Change Agent to clarify and agree on mutual expectations with colleagues is to treat them from the outset as internal customers. This signals a commitment to serving their needs, and provides the opportunity to demonstrate methods and tools for communicating with external customers.

Step one is to clarify needs. Sit down with individual team members and with their help, write down what is important to them, what they need from you, and what they don't need. (You should prepare beforehand a list of your own ideas regarding what you think they need.) Then ask them to assign priorities, because you may not be able to do everything they would like.

Step two is to analyze the wants of all your customers, and to figure out what course of action will provide the most value to all of them, while helping the organization meet its goals.

Step three is to negotiate, agree on realistic expectations, and commit to these. You should be flexible about directing effort to where your customers perceive the most value, and firm about not committing to more than you can do.

Step four is to gather feedback regularly and systematically on whether you are meeting their expectations, and to update your commitments and their expectations, based upon their changing needs and your changing capabilities.

How was this process mutual? When making commitments to others, you will also need to get commitments from them to do their part—the customer always has some responsibilities to fulfill in order to make the supplier's job possible.

- How much personal involvement do other senior people (especially the leader) expect to have? What forms do they expect this to take?
- What role do they expect you to play? Can they explain this with any precision?
- What tangible responsibilities will your colleagues have for the success of the change process? Will these be part of their personal objectives?

- Who besides the president knows about the plans for change and has given their support (for example, has the board given approval)?

More often than not, in an organization that is just embarking on a formal improvement process, many of these questions are unanswerable. You may find that your colleagues have very different and contradictory ideas regarding the goals of the organization and what they expect of you. As far as your role is concerned, you may not like some of their ideas at all!

Your colleagues will probably be looking to you to make proposals, for example, regarding objectives for the change process, the contribution that you (and they) should be expected to make, and so on. This is not an unusual situation—in fact, this is to be expected. Your task is to manage the discussion, offer suggestions and proposals, point out contradictions, and work toward a common understanding.

This process will not happen overnight, and it will never be complete. Your shared understanding of the goals and the tasks involved will evolve over time, and so will your role and your objectives. As this happens, be sure to capture the key areas of agreement. This information will be valuable as you begin to develop more detailed plans for the transformation.

Agreements with the leader and the team

You should ensure that you and your leader establish a clear shared understanding of what you are trying to accomplish, your immediate (joint) goals, what your respective roles will be, and the mechanics of how you will operate.

Exhibit 3.8 offers suggestions on the president's role in supporting the process. Chapter 1, "The Job and the Person," also provides background reading for *both* of you on the role of a Change Agent.

The following are typical agreements for which you might strive with the president and your colleagues. You will not necessarily secure agreement on all of these.

- You will have regular one-on-one meetings to discuss progress and problems and to plan ahead.

• You will each write out and agree to your respective roles and responsibilities for the change process. You will share these with the team.

• You will jointly figure out how to establish and develop a top-level management forum to plan and oversee the change process (we will refer to this forum as the *Quality Council*).

The president will chair the meetings of the Quality Council and hold all the members accountable for meeting their commitments. You will help plan and facilitate the meetings.

• Each senior manager will allocate a certain amount of time to support the process. You will provide ideas and suggestions regarding how best to use this time (for example, for education, visiting customers or other leading companies, small-scale process improvement projects, and so on).

• You will set up some mechanisms to provide all of the team with feedback on whether everyone is perceived to be "walking the talk." As a result of this feedback, there will be a desire for coaching to help senior people behave in a manner consistent with the quality approach. Your colleagues will see you as the natural person to provide this. This is a sensitive area, which requires a high level of mutual trust. It's only fair and sensible to make this coaching a shared responsibility and to expect your colleagues to coach you, too.

You may want to make it clear right from the start that you view such coaching as one of your responsibilities. However, even when you have such a mandate, you need to choose carefully the timing and the tone of any unsolicited coaching, or this may be a career-limiting move! We're all human, and no one likes to feel that they're being kicked when they're down.

• You will challenge the team when its commitment seems to be flagging or when the process seems to be losing momentum. For example, if the team is forgetting to take into account the impact of decisions on customers, it is your job to speak up. You might want to point out in advance that the team will not always appreciate your input at the time; but, if they want it, they must refrain from "killing the messenger."

Tangible goals and objectives

The following are typical objectives that you might agree on during the first 18 months or so.

Exhibit 3.8 *The president's responsibilities for the change process.*

- Develop a consensus among the top management team members regarding where they want to go in the long term and how they can get there. The quality journey should be an integral part of a broader vision.
- Oversee the development of a detailed plan for the journey, including tangible objectives for every member of the top management team.
- Play an active part in communicating the plan throughout the organization, and monitoring understanding and reactions.
- Review progress against the plan at regular intervals, and hold members of the management team accountable for their commitments.
- Analyze outcomes with a view to understanding the cause and effect relationship between actions planned and the results.
- Identify and recognize valiant efforts as well as successes, both formally and informally.
- Establish cooperation and teamwork among top management as the normal mode of operation.
- Sponsor the continuing education and development of the top management team. Typical issues might be participative management styles, process improvement, and the use of the planning tools (see Appendix C).
- Act as a role model for attitude and behavior, take a personal interest in progress at a grass roots level, and maintain direct personal contact with customers.
- After allowing a reasonable time for learning and adjustment, take firm action to minimize the negative impact of senior people who will not or cannot buy into the process.

- You will orchestrate the development of a start-up plan with input from all, including estimates of the budget and other resources required to execute this.
- You will support the president in the task of making the top management team more cohesive (assuming that this seems

necessary) by offering ideas and guidance regarding possible methods.

- You will project-manage some elements of the plan, such as the procurement of education.
- You will provide reasonable assistance to colleagues, on request, in executing their parts of the plan—without relieving them of their responsibility to meet their commitments.
- You will work with the top management team members to identify and provide educational opportunities suitable for them.
- You will assist your colleagues in identifying and developing suitable people within their functions to support the change process.
- You will help identify suitable external resources to support the plan; for example, providers of education and training materials, or experts in specific areas such as survey design.
- You will facilitate (but not take ownership of) at least one process-improvement or problem-solving team, with the aim of securing some early tangible benefits.

Setting realistic objectives

The big issue here is "How long will it take?" Often your colleagues may believe (or hope) that bottom-line results will come quickly—and you might also like to believe this. If only it were true—a dash of training, a pinch of motivational speeches, a teaspoonful of teamwork, and sales improve miraculously, costs begin to melt away. . . . However, back to reality.

One of your tasks is to manage expectations about how quickly the team can expect to see progress and results. The following is an approximate calendar of typical events.

Building up steam The first phase is for top management to spend time getting educated, developing a consensus that this is the way to go, considering its approach and developing an initial plan for the transformation. During this phase, the first assessment of the current status also should be conducted to provide a basis for creating the plan.

The initial plan may be quite modest in scope. What makes it the starting point is that the ultimate goal is the transformation of the organization.

There is no typical time frame for this phase—it may take months or years.

Starting execution The initial plan is agreed upon, with target areas for results identified; the Change Agent and trainers/facilitators are in place; and top management is meeting regularly to review progress.

Six months Targeted education for early participants is well under way. Some problem-solving and/or process-improvement teams have been formed and trained.

One year Some teams begin to report significant progress. Some project work is nearing completion, such as the new customer-satisfaction survey. The annual report on the process shows many activities completed and under way, but few tangible results.

Eighteen months A few teams have completed a first cycle, others are still in progress or stuck. The teams that have completed their task have fixed some causes of recognized problems and identified ways of tackling other causes. A few teams can demonstrate significant bottom-line dollar savings. Some other teams have results that seem significant, but the benefit is not easily translated into dollars (for example, measurably improved customer service).

Some of the most enthusiastic people are becoming weary, frustrated that many others are not doing anything. Meanwhile, those who oppose the change are making a last stand, arguing that the changes are not accomplishing much and are not worth all the effort involved. "Bring back the good old days," they are saying.

Two years A significant change is apparent in the style of working in many parts of the organization. There are many small achievements that have been recognized and publicized, and a few that have significant bottom-line impact.

At the end of 18 to 24 months, it should be clear that the collective efforts of the management team have created some change in the desired direction. There should be

- Significant, noticeable changes for the better in behavior in some parts of the organization
- Examples of tangible benefits flowing from problem-solving and process-improvement efforts
- Evidence that enables management to project a significant impact (for example, on productivity or customer satisfaction) from improvement trends being sustained and improvement efforts being pursued on a larger scale

In some cases, there may already be significant impacts visible on operating efficiencies, such as greatly reduced inventories, greatly reduced cycle times, greatly reduced errors and waste. These are the "low hanging fruit." However, such major tangible gains will only materialize if, in the very early stages, the management team was able to identify where such gains could be made and the actions required to secure them.

If it was not clear at the outset how execution of the plan would lead to tangible results, do not be surprised if you don't get any.

The actual time frame will vary greatly according to the size and complexity of the organization. One way of thinking about this is that the typical unit of improvement is often a group of approximately 50–200 people who have some common purpose. In a large organization many such groups typically spring up, in divisions and in corporate functions. This calendar may be typical of such groups, but it may not be typical of the organization as a whole.

This information is provided simply to help you avoid raising unrealistic expectations at the outset regarding time frames. When you have completed your first detailed plan for change, you will be in a much better position to estimate the nature and timing of the outcomes in your particular situation.

Behavioral goals

In additional to the tangible goals, you may also want to set some goals that are somewhat less concrete, such as changes of attitude and behavior within the management team. There are good reasons for doing this.

- Some of your key personal achievements will be of this nature, and you need goals like this to motivate yourself.
- These changes are "enablers" for other more measurable achievements of the organization.
- These changes are the first signs of progress.

It is worth discussing the idea of such early indicators, especially with the president. Some behavioral changes are difficult to quantify, but you can choose simply to be on the lookout for incidents that signal changes in the culture.

For example, your collective efforts will have had a major impact on the organization if

- When making decisions, the management team usually remembers to consider the potential impact on customers.
- Management views employee morale as an important issue and shows as much interest in the employee survey as in the sales figures or the stock price.
- Managers treat others in the organization with equal respect regardless of their status.
- The most territorial member of the team is willing on occasion to volunteer shortcomings in his or her operations rather than becoming defensive or blaming others.
- The most impatient and hot-tempered member of the team has developed some restraint and listening skills.
- The most withdrawn member of the management team has begun to speak up and to offer thoughtful opinions.
- The most intuitive, "shoot from the hip" member of the team has been seen to draw a Pareto chart, or to ask for hard data about the possible causes of a problem.
- When an organizational goal is endangered by problems in one department, some other top managers offer genuine

support and assistance to their colleague and have their own departments bear the cost of this support.
- Employees and middle managers who report on their efforts to the Quality Council are less nervous in this situation. They are more outspoken and candid.
- And so on. . . .

You and your colleagues should look out for and recognize these changes as they begin to appear. These are often the first signs that your efforts are really succeeding. Equally, a drop-off in these small signals may be the first sign that not all is well.

You should all be prepared to challenge each others' behavior when it seems inconsistent with the organization's declared principles. We will discuss later how you can begin to work toward agreement on what the principles—and, hence, these behavioral goals—should be.

STEPPING UP TO THE NEW ROLE

Personal education and development—achieving balance

There is a story of several blind people describing an elephant. One, grasping the tail, was certain that an elephant was like a snake; another, grasping a leg, was equally sure that an elephant was like a tree; and so on.

You must avoid the trap of seeing only a part of the whole. This is most important, because a quality approach is like a machine with several critical components. Neglect any one of them, and the whole system will malfunction.

This book is devoted to setting out the full breadth of what you need to know, without necessarily going into great detail in any one area. It gives you the "big picture"—and hence an idea of the range of skills and knowledge that you should aim to acquire.

To achieve this type of broad understanding usually requires effort in areas to which you may not be particularly attracted. Your past experience and training may have sharpened your

Exhibit 3.9 *Common professional strengths and development opportunities.*

	Human resources	Marketing	Quality Assurance	Operations
Common strengths	• Focus on relationships and human issues • Training and facilitation skills	• External customer focus • Selling of ideas—by focusing on opportunities not deficiencies	• System thinking and understanding of processes • Use of data and statistics • Organized, methodical approach	• Focus on outputs/results • Use of data • Attention to detail
Common development opportunities	• External customer focus • System thinking and understanding of processes	• Understanding of processes • Promotion of team efforts as well as individual accomplishments	• Expertise in soft issues • Selling skills • Pursuit of opportunities rather than deficiencies	• External customer focus • Expertise in soft issues • Working through others without direct authority

focus in some ways, yet made it harder for you to see a different perspective.

Exhibit 3.9 sets out some common strengths (and development needs) that are often associated with training and experience in certain disciplines. These are generalizations that don't apply to everyone, and every individual is different. But, it may be worth considering whether the mind-set fostered by your training tends to blind you to other important issues. These are key areas for your own personal development.

Another reason for delving into how your own mind works is that you need to explain and discuss the concepts of quality with people who may think and work in a very different fashion from you. If you are strongly left brained by aptitude and by training (for example, an engineer), you may naturally tend to think of quality in logical and mechanistic terms; while your more right-brained colleague (for example, a psychologist) wants to understand the subject more in terms of attitudes, emotions, and relationships.

You will be much more successful as an advocate and as an educator if you can understand how others think about and understand the world—and this may be very different from your own preferred thought process. They won't change—why should they? But you can communicate with them if you will learn their language.

Working at a higher level

It is not unusual for the appointment of a change agent to be viewed as a development opportunity for a promising person within the organization. If this is you, congratulations!

This is a splendid opportunity for you—a wonderful learning experience, a chance to make a difference to the organization, and a chance to work and become visible at a higher level in the organization.

It is also a tough developmental assignment, and because this is a new role, it will pose more of a challenge than simply taking over one of the existing top management positions.

If you are full of confidence about working at this level, feel free to skip this section. However, if you feel that you could use a few pointers, then don't be too proud to examine the following ideas.

A mismatch of expectations?
First of all, let's not shy away from one possibility that you might be tempted not to consider. If there seems to be a large gap between yourself and your new colleagues—in terms of experience, previous job level, and so on—perhaps you should test more carefully what role they expect you to play. It is possible that their expectations are too low.

For this process to work, there must be a champion for change within the team who is viewed as a full member and a peer, who can challenge colleagues without giving offense, and who is expected to contribute to all top management decisions. If your colleagues expect a more modest contribution—perhaps just some project-management support—then there's a mismatch of expectations that you need to address.

You could test the situation by exploring more precisely your colleagues' expectations—especially those of the president. If it is clear that they expect or want something less than a fully fledged peer within the team, then you need to make a judgment regarding whether the situation holds the potential for success. The issue is not your ability—it is the role you are cast in by your colleagues and the amount of experience you can bring to the task.

If you believe that you can successfully break out of this limited role, that's one way forward. Another might be to identify at least one sponsor within the team who is strongly committed to the change and is prepared to be your mentor and supporter. The Change Agent's role is then effectively shared between you.

You may have other ideas. The only firm rule is: try to find an approach that is likely to work. If you can, then go for it. If not, spare yourself a lot of needless pain.

Working with senior management
The only generalization that one can make about senior management is that one can't generalize about senior management. As in all human endeavors, there is enormous variation in how management teams operate, ranging from awe-inspiring to awful. So the first suggestion is

- *Don't be over-awed or blinded by preconceptions.* You discover that your new colleagues never use meeting agendas or record what they have agreed in meetings? Do not conclude that top managers are so clever that they can dispense with such tools.

As one goes higher in the hierarchy, the level of *individual* competence, talent, and experience generally rises. But, the level of competence *as a team* sometimes declines. This is usually due to intense pressures on the individual—pressures to meet performance targets, pressures to compete in order to meet conflicting objectives—all fostered by the system in which they work. So don't be surprised if teamwork isn't really the norm within the top management team, or if it is difficult to get the team to focus on longer-term issues.

- *Remember that top managers are only human.* However formidable their demeanor or their talents, there is an ordinary, vulnerable, well-intentioned human being under the skin of every senior manager. At home they're just Jim or Ann, scolded by their spouses when they don't clean up, concerned about their kids, fearful about what their next medical checkup might reveal.
- *Demonstrate complete commitment to the task.* Be prepared to blitz the task up front. Getting up to speed quickly will make your job easier subsequently, and allow you to get back to a more sane lifestyle. Always be prepared. Do your homework.
- *Become attuned to the working style of your colleagues.* The working style of your colleagues will usually be more strategic than operational, more concerned with the big picture than with details. Their ways of communicating may also differ greatly from what you have been used to. For example, many people at senior levels don't telegraph their feelings and intentions—they may need to be adept politicians in order to survive. You're more likely to hear "I wonder if this is really the best course of action," rather than "You're talking nonsense." On the other hand, some management meetings look and sound like barroom brawls. These have the benefit that you can easily tell who is on which side.
- *Listen to your instincts and stick to your principles.* Your reputation for integrity is one of your main assets. Do not be intimidated into surrendering it.
- *Be prepared to establish your right to be there and to be heard.* Weakness or lack of resolve is never an asset in these circles.
- *Be prepared to admit ignorance and ask for help from your colleagues.* You might feel strange and unusual asking for help, but that's OK. You are going to do even more strange and unusual things.
- *Use your information network and your personal support network.* Your information network is comprised of peers in other

organizations who can provide technical information. Your personal support network is a circle of friends who provide moral support. Chapter 8, "External Resources," explains how to establish and use these. For advice about how to work at a higher level, use your network to seek out an experienced mentor, probably outside the organization, and draw upon his or her wisdom.

As you get to know your colleagues better, and as your confidence grows that you know your way around the organization, you can become more focused on the task in hand—helping the team to develop a plan for the transformation.

4

Preparing to Launch the Process

There is a time for departure even when there's no certain place to go.

—Tennessee Williams

CHAPTER CONTENTS

Moving from a commitment in principle to a detailed plan of action.

- Building and sustaining the initial commitment
- Developing a cohesive top management team
- Establishing and developing a Quality Council
- Reestablishing the mission, vision, and values
- Conducting an initial assessment

This chapter deals with the preparations and planning that lead up to the launch of the change process. This phase starts when a compelling reason for change has crystallized, and top management has decided to act. This phase may be considered complete when top management has fully committed itself to a major transformation, completed development of the first detailed plan for change, and begun execution of the plan.

Once top management has decided that there is a need for major change and made a commitment in principle (and designated a Change Agent to help), the next stage is for the team to

- Figure out what kind of change is really needed. What exactly is management trying to accomplish, besides making everything "work better"?
- Figure out how this could be accomplished.
- Develop a detailed plan.

This phase is *preparing to launch the process.* This should be much more predictable than the previous phase (securing an initial commitment) because a decision in principle has been taken and a Change Agent appointed with a mandate.

However, executing this phase is still not straightforward. There is much to do, and much uncertainty at first about where this is all leading. The initial realization that something must be done needs to be reinforced, channeled, and translated into a plan for purposeful and concrete action.

This phase is rather like the discussion that takes place between an agreement in principle ("Yes, it's time for a relaxing family vacation") and an agreement on the specifics ("We'll go to a condo in Barbados for two weeks during May"). During this period, the nature of the vacation must be agreed upon, and all kinds of different personal priorities must be reconciled, ranging from water sports and nightlife for the adventurous, to serene and scenic surroundings for those who are more reflective. There may be squabbles when some individuals feel that their needs are not being considered. In the end, if there is goodwill and a desire to make it happen, a plan is agreed upon that everyone can live with. Then we have to deal with the fact that some people just will not start packing until an hour before the plane leaves.

Exhibit 4.1 *Typical launch preparation activities.*

Activity	Issues
• Establishing a team (the Quality Council) to oversee the change process	• How do we organize ourselves to plan and oversee the changes?
• Reestablishing the mission, vision, and values	• What business are we in? • Where do we want to be? • What principles do we adhere to?
• Conducting an assessment	• Where are we now?
• Developing a plan	• How do we get from where we are to where we want to be?

This preparation phase is complete when

- Top management has clarified its *goals* and the *principles* that underlie the team's strategy for getting there.
- Top management has developed a *road map* for the journey and a *detailed plan for the early steps.*
- Top management *owns the plan* and is committed to play an *active* part and to *lead by example.*
- Top management has, in the process of developing the plan, become educated sufficiently in order to arrive at an *informed commitment* to embark upon the transformation *in this particular way.*

Exhibit 4.1 shows the typical activities involved in reaching this outcome. Before examining these next steps, let's examine some ideas that we need to bear in mind throughout the process. These are advocacy for the process, and the development of a more cohesive top management team.

Advocacy

Now that a commitment has been secured, does this mean that we can switch our attention to implementation and forget about advocacy? The answer is definitely not! The initial commitment is always weak in the sense that it is mainly intellectual, and those involved have not yet had exposure to the practical consequences of their decision. The team may have second thoughts as the magnitude of the task becomes clear.

The initial decision still has to be translated into emotional commitment and personal involvement. As the plan is developed and deployed, there is a need for advocacy to sustain commitment through both good and difficult times. When some success is achieved, advocacy is still needed to combat complacency and to build the commitment needed to continue.

So, you should not think of the initial commitment as an irrevocable, long-term license for change. Think of it more as a temporary permit that has to be renewed at frequent intervals. This permit is prone to expire whenever top management expectations are not being met, when difficulties are being encountered, or when there is a feeling that the job is complete. Advocacy for the process is never done.

DEVELOPING A COHESIVE TOP MANAGEMENT TEAM

The people are a reflection of the management.

—Phil Crosby

We have been referring to the top management group as a *team*, but often the current reality is very different—to the great regret of the president and everyone else. Often, divisive forces—internal competition, intense pressure to achieve departmental goals, and unintentional reward of self-serving behaviors—have seriously undermined the forces that can help bind a team together. These binding forces include

- Shared goals
- A challenge or an external threat ("We may not survive")

- A recognition of interdependence and, hence, a desire for cooperation
- Mutual trust and respect

When the divisive forces are predominant, teamwork becomes impossible, and members of the group often have to put their own interests before those of the organization, or be the losers. This is no fun, and it is not an ideal state of affairs, but it is not unusual. Is this really a serious concern?

If top management does not function as a cohesive team, this presents a serious barrier to implementing a quality approach because *the rest of the organization reflects the top management team. It cannot be better.* This is especially true for the issues that the organization now has to work on. Attitudes and behaviors propagate throughout an organization, inexorably led by the example of top management.

- Frontline people will never become enthusiastic or involved in making things better until they are treated as thinking people who want to do a good job, and whose work is important to someone.
- Customers will never be treated with real concern and respect until the people who deal with customers receive similar treatment from their supervisors.
- The departmental barriers that impede the flow of work cannot be broken down—and the key operating processes will never work well—until the departmental leaders learn to cooperate well with each other.
- The patient, low-key work of getting to the root of problems and preventing recurrences will never become widespread until top management values this work more than spectacular fire fighting.

For all these reasons, a key task in changing the behavior of the organization is to change the behavior of the top management team.

This will take time, and it cannot be tackled as a separate step or as a precursor to further action. It has to be accomplished in the course of getting the job done; that is, during the team's efforts to plan and support the transformation. Teamwork needs to be put on the table as an issue, but it needs to be worked on as a means to an end, not as an end in itself.

Exhibit 4.2 *Eleven commandments for an enthusiastic team.*

1. Help each other be right, not wrong.
2. Look for ways to make new ideas work, not for reasons they won't.
3. If in doubt, check it out! Don't make negative assumptions about each other.
4. Help each other win and take pride in each other's victories.
5. Speak positively about each other and about your organization at every opportunity.
6. Maintain a positive mental attitude no matter what the circumstances.
7. Act with initiative and courage as if it all depends on you.
8. Do everything with enthusiasm. It's contagious.
9. Whatever you want, give it away.
10. Don't lose faith. Never give up.
11. Have fun!

So, most of these early steps, while aimed at creating some tangible output, should also be used as opportunities to introduce ideas and methods that facilitate teamwork. These ideas will be welcomed because they make the task easier, and because a true team effort is more enjoyable and more effective. In this way, the top management team will become more cohesive and better able to model and support the desired changes.

ESTABLISHING AND DEVELOPING A QUALITY COUNCIL

As you get to know your colleagues better, and understand how they operate, you will feel more comfortable about working with the team in a group setting. This is where a lot of action will now take place.

In order to plan and lead the change process, top management must set aside adequate time to work on this task as a team. Some substantial off-site meetings will be called for—for education purposes and to work through major decisions together—as well as regular meetings to review progress and to make routine decisions.

Rather than schedule this work in an ad hoc fashion, or attempt to combine it with other scheduled meetings that have very different aims, it usually makes sense to give this team effort an identity by giving the team a title, such as "The Quality Council." Doing so also conveys the idea that this team will continue to operate for some time to come.

The Quality Council's mandate is to improve the way that the organization is run, using quality management principles and practices, so that the organization becomes more capable of achieving its goals. This mandate covers every department and every aspect of the organization, so a more descriptive title might be "Council for the Improvement of the Entire Organization." However, the council's role is developmental, not operational.

The Quality Council membership should include all those who are key decision makers in the organization. It may also include some others who represent stakeholder groups. For example, the council may decide that it needs to involve representatives of staff associations or unions. One important test of the chosen membership is that the council should have the same level of decision-making authority as any other formal meeting of the top management team. It should not have to seek top management approval for its decisions, because it *is* top management.

Working with the Quality Council

While you were settling into your new role, many of your dealings with your colleagues were one-on-one. There are some key distinctions between discussions and agreements which take place with your peers in one-on-one meetings and those which take place as a management group.

- The team has a larger mandate. It is (or it should be) concerned about the organization as a whole, not just individual departments.

- The team's decisions bind the individual members.
- Commitments made in these formal meetings are commitments to the entire team.

So all of the vital decisions and agreements regarding the change process must be made (or finally approved) by the Quality Council, recorded, and treated as commitments to be followed up on in the same way as other formal decisions made by top management.

Here are a few issues that you may want to get on the agenda during early meetings.

• The mandate and objectives of the Quality Council, the schedule for future meetings, typical agenda items, and roles in the meeting (for example, chairperson and planner/facilitator).

• Meeting ground rules. It may help to agree to some ground rules that will enable meetings to be more constructive and productive. This is also a step toward better teamwork. (With some groups it may be vital to do this early rather than later.) For example, the telling of lengthy war stories may be banned by mutual agreement, or the practice of "killing the messenger"—that is, attacking those who bring bad news or facts that are displeasing—may be outlawed. Exhibit 4.3 describes one way of establishing some meeting ground rules.

When ground rules are agreed upon, it is a good idea to post them visibly and to agree to evaluate each meeting against them as the final agenda item. *Remember, a commitment without a follow-up mechanism is often worthless. Always look for a way to close the loop.*

• The planning process. It does not make any sense simply to dive straight into debates regarding how best to launch the process, or to accept delegation of the entire task onto the shoulders of the Change Agent. Planning the launch is a major task that should be tackled methodically and involve everyone on the team.

You should first agree with your colleagues how this planning should be carried out, who needs to be involved, and so on. It is fair for the team to expect you to put forward ideas and proposals.

You have an opportunity here to make use of some appropriate tools. For example, the process for planning the launch can be set out in a simple flowchart; brainstorming can help to identify ideas for action, and these can be grouped into themes using the affinity

Exhibit 4.3 *Establishing behavioral ground rules.*

Here is one way to establish behavioral ground rules.

- Arrange a management session to fill in the blanks in the following statements:

  "Around here, when it comes to ______, the way we work is _____."

  "Around here, when it comes to ______, the way we *will* work is _____."

For example,

"Around here, when it comes to *decision making,* the way we work is *the president decides and we do as we're told.*"

"Around here, when it comes to *decision making,* the way we *will* work is *to arrive at a consensus by open discussion of the facts.*"

- Repeat this with a few behavioral issues—those that the members of the team consider important.
- Use the second statements as the basis for posted guidelines.

diagram process. This approach may facilitate the task and give you and your colleagues some practice if these tools are new to you.

Educating the Quality Council

At this early stage, there may be a need to raise awareness and understanding without implying that your colleagues are unaware or ignorant today. This initial gentle education will usually lead to a realization that there is much more to be learned and a desire for more formal education. Here are some ideas for how to start this awareness-raising process.

- Create situations in which people learn from an experience. For example, direct contact with customers or employees

Exhibit 4.4 *Do senior managers need to understand the "nuts and bolts" of quality improvement?*

Senior managers do not absolutely need a detailed understanding at the outset of the methodologies, tools, and techniques, but they need to acquire some of this knowledge at some stage in order to be effective.

Initially, managers only need to know enough to understand that a quality approach will help them to be successful. However, they should master some of the basics as soon as possible, for two important reasons:

- To equip them to improve their own critical management processes (like the planning and budgeting process)
- So that they understand what their own people are doing and can support these efforts

When top managers are acting as highly visible advocates for quality, they need to understand why, for example, they may see frontline people working with Pareto charts or cause-and-effect diagrams. Rather than mistakenly chiding employees for not "getting on with their work," senior people should immediately realize that some problem solving or process improvement is under way, show an interest, and offer encouragement. These employees are prospecting for gold!

can reinforce the need to understand better their needs and concerns.

- Educate by example. You have countless opportunities for modeling new ways of doing things—in the way you establish the requirements when undertaking projects to support your colleagues; in the way you organize these projects; in the way you obtain feedback from your internal customers upon completion; in the way you collect, display, and share data; in the way in which you work with the people who report to you.
- Look for ways to tackle current top management tasks more effectively using the quality methods and tools. If top

management is dissatisfied with the planning and budgeting process, you could suggest flowcharting the process and using this as the starting point for some problem solving.

Making it fun

It will help the whole organization if the top management team members can have some fun as they begin to work together on improving quality. One of the most deadening beliefs prevalent in organizations is an unspoken assumption that people cannot be doing good work if they are having fun. Laughter is thought to be a sign of "goofing off." Seriousness—a commitment to the shared goal—is confused with solemnity. The reality is that humor is one of the most valuable ingredients in building a productive working environment, because

- It is part of the process of building relationships.
- It is a tool for feeding imaginative and creative thought processes.
- It is a vehicle for communication and education. Sometimes we can only understand the foolishness of how we do things when we see the funny side of this.
- It diffuses anger and relieves stress.
- It is healing.

A humorless environment is characteristic of dictatorships and totalitarian societies where everything the leader says is supposed to be profound and infallible, and no one is allowed to poke fun at the foibles and excesses of those in power. We don't tolerate this type of thought censorship in our society—why should it be put up with at work?

The reality is that humor cannot be completely suppressed. When it is not appreciated it simply goes underground, and it is then perceived as a form of subversion. *When the inherent human desire to laugh and have fun has gone underground, this is a sure sign that the hierarchy of command is smothering the organization—suppressing information, killing initiative, feeding the egos of those in power.*

So, encouraging some fun at work seems like a smart idea. If the working environment in your organization resembles that of a funeral, there could be a lot of positive energy waiting to be released if this draining atmosphere can be changed.

But how on earth can this be achieved? Will a "fun committee" of top executives be required to examine ways of changing the culture? Should consultants be called in? How much will it cost? Will the bottom-line benefits justify the cost? What will it do for the stock price? It's really quite simple.

- All that people need is permission. People will naturally have fun at work if the barriers to normal human behavior are removed. *Normal human behavior* includes wanting to do a good job and feeling proud about it, wanting respect from and positive relationships with others, wanting to celebrate success, wanting to mourn briefly and put setbacks in perspective, and wanting to relax and recharge from time to time.
- It is easy to create opportunities that show that permission has been granted. Brainstorming, for example, is most creative when it is a wild affair with ridiculous ideas being put forward. These provide the seeds for innovative but practical ideas. The facilitator can set the tone in this and other team activities. And, if top managers take themselves less seriously, others will notice and stop pretending.
- A bit of planning can help, too. There should be planned opportunities for celebration and relaxation—designed with help from the people who will take part.

The evolution of the Quality Council

This section deals with the issue of where and how management should spend time working on quality, and how this may evolve over time. This may seem like a long-term issue, but it will help you right from the start to know where you are going in the future.

There are two ways for management to allocate time to working on quality: (1) to add *quality* (or *improvement,* or whatever) to the agenda of the existing management meetings, or (2) to convene another meeting of essentially the same people, but with a different agenda. Both of these approaches have benefits and drawbacks, as indicated in Exhibit 4.5.

Exhibit 4.5 *Integrating quality management into the organization.*

Establishing a separate meeting schedule

+ Ensures that top management does spend time on the issues.
+ Provides a clean slate for establishing a mode of operation, behavioral ground rules, and so on.
+ Permits opening up of this forum to other stakeholders, for example, unions.
– Increases the risk of becoming disconnected from the organization's objectives and from the normal operations and structure.

Adding quality issues to the agenda of other regular management meeting

+ Ensures integration with the normal management structure and decision making—an important goal.
– Increases the risk of quality "falling off the bottom of the agenda"—no time or energy left.
– May cause a possible clash with normal agenda, or a clash of mind-sets or time frame—the group cannot move instantly from reviewing day-to-day problems and results, to thinking about how to improve the organization as a system.
– May not signal to the organization that anything has changed.

The approach described in this book combines a separate meeting (the Quality Council) with a linkage to other established meetings (such as the operations committee). This may not be the arrangement chosen by your team, and it is not put forward as the one right way, but it will serve to illustrate the issues involved.

In the long term, the aim is to have quality improvement activities seamlessly integrated into every aspect of running the organization. Then there may be no need for a Quality Council. In the meantime, it is essential to define the relationship between the Quality Council and other top management teams or committees. One way of looking at this relationship is that the Quality Council works to develop and extend the management system, but does not operate it.

A revised measurement system would be a good example of this. The Quality Council might devise the new measurement system and might also project-manage the development work, but the completed system would then become the responsibility of the operations committee. This group would take responsibility for reviewing the resulting measurements on a regular basis, and for responding to out-of-line operational situations revealed by the measurements.

Another good example might be the planning process. One of the major functions of the Quality Council is planning for ongoing improvement. However, if the Quality Council can develop a planning process that encompasses the various needs of the organization, including the need to plan for improvement, then this aspect of the Quality Council's work can be subsumed into the normal planning cycle. In this way, the Quality Council may, over time, work itself out of a job by achieving full integration of all quality improvement activities.

REESTABLISHING THE MISSION, VISION, AND VALUES

Strip away the trendy buzzwords, and a fundamental truth remains: the organization has a much greater chance of success if it is clear about what business it is in, where it wants to be in the future, and what operating principles are important to it. It will have even more chance of success if these are not secrets held by top management, but are understood by everyone.

When the mission, vision, and values are undefined, are not communicated, or are not real or inspiring to people, then this can undermine any efforts to improve. How can you identify the target customers you intend to delight if you are not sure what business you are in? How can you develop any kind of long-term plans to stay in business, if you have no idea where you want to be?

People already know what behavior is acceptable and unacceptable in the organization today—by observing what gets rewarded or punished. But are these the types of behavior you need to make this a great organization?

If the organization is in poor shape in any of these areas, then work will be needed before other positive changes can be achieved. This process must start with top management. Only when top management is united on these issues can it provide direction to the rest of the organization. Communicating the vision and values throughout the organization is a challenging task—but until management shares, believes in and lives the mission, vision, and values, it is an impossible task.

Building a consensus around these issues—mission, vision, and values—is an important step toward creating a cohesive management team. This presents another opportunity for you to help—by suggesting processes that will enable the top management team to achieve this consensus.

Mission

Definitions vary greatly, but here we will take *mission* to mean *what business we are in.* Therefore, mission defines purpose and scope. A mission statement should help the organization focus its efforts within certain boundaries, yet allow for evolution of its approach as times change. "Making a profit" is too broad a mission statement, and "manufacturing ice cream" might be too narrow. "Supplying specialty foods to retailers" might be about right. Sometimes the mission is set by an externally determined mandate; for example, in the case of a government department. When considering what business you are in, it may help to examine what are the organization's core competencies, which are closely related to its key processes. Hopefully, the team can readily agree on the purpose of the organization, and move on to vision and values, which are usually harder to develop.

Vision

By *vision* we mean the definition of *what we want to become.* Vision defines the desired future state of the organization, and, hence, its long-term goals.

To develop a shared vision, the team should work through a well-designed visioning process. This is a team-building process, as well as a starting point for strategic planning. This process will draw out the diverse individual perspectives and priorities within the team, and help establish common ground. This vision should be developed sufficiently to suggest a few key priorities for the whole organization and, thus, provide some inputs to the planning process. (An assessment of the current status will provide the other main input.)

There are texts describing the value of a visioning approach[2] and video material that can be used to support the process.[3] You may wish to engage an experienced external facilitator to lead the team through this exercise.

Values

Values are like operating principles, which should guide all decisions and actions. These typically deal with issues like integrity, respect for people, and service to customers. It is best if these principles can be kept few in number and easy to recall.

What is the purpose of discussing and agreeing on these? There is no purpose at all unless they will be used. Sometimes considerable management effort is expended on discussing and agreeing on these words, then announcements are made, reminder cards and posters printed—and that's it! There is nothing wrong with these actions, but if nothing else is done, there are some important elements missing.

- There is no method of putting the principles into practice.
- There is no mechanism for closing the loop; that is, for determining whether these statements are becoming accepted as the guiding principles for behavior. If not, the old unwritten rules will remain foremost in people's minds—don't rock the boat, worry about the customer after you've made your numbers, and so on.
- Most people in the organization have had no say, and never will have. Why should they embrace and live by principles that they have had no say in determining?

To make the values operational, the team might resolve to use the declared values as a screen for all their important decisions. What about everyone else in the organization? These others are not the problem. If the values are made real within the top management team, everyone else will soon notice and begin to pay attention. (See Exhibit 4.6, "Making the values operational.")

Exhibit 4.6 *Making the values operational (a true story).*

Soon after we drew up our values, we had to decide what to do about the cost of personal long distance phone calls from company telephones. These costs were getting out of hand, and we had obtained a quote from a supplier for installation of a call-monitoring system. This was not cheap, but it was estimated that this system would trap virtually all of these calls.

However, after discussion, we felt that this approach conflicted with our first value: respect for people. What should we do? In the end, the president sent out a letter to all employees explaining the problem, and we set up a very simple procedure for employees to identify personal calls so that they could reimburse the company.

The result was that the cost of these calls to the company dropped by 90 percent and we did not have to incur the cost of the monitoring equipment. Best of all, we avoided sending a message to our people that we didn't trust them to act responsibly.

To confirm whether the values are being respected the team might include questions regarding these in various feedback systems (upward appraisal, 360 degree feedback, employee surveys, and so on). Again, the initial focus should be on the top management team. These are the people whose lead others will follow, who most need the feedback to ensure that they are "walking the talk," and who don't get such feedback in the normal course of events.

There is no limit to the time that can be spent debating mission, vision, and values. However, where these foundations are weak, there is a great deal of benefit in rebuilding them. This process takes time and can be frustrating, especially if the team feels that there are other urgent priorities.

Your task is to help the team decide how much effort is needed in these areas, and to devise an approach that gets the job done while making the best possible use of everyone's time. If it can be fun too, that's a big bonus. If this work is done well, there may already be some significant changes evident in the way that the team functions, due to

- The common ground set out in the mission, vision, and values
- Personal contact and improved understanding of other people's perspectives
- The experience of working together in a constructive manner

CONDUCTING AN EARLY ASSESSMENT

Having figured out where the organization wants to go, it now makes sense to take a look at where it is now. Your next task, on behalf of the top management team, is to orchestrate an assessment.

An *assessment* is a kind of self-audit in which the organization compares itself to a set of criteria that represent an ideal. It is like holding up a mirror in which senior managers as a group can study themselves and their handiwork. They may not like what they see, but if they accept the accuracy of the image, they will feel compelled to react to it.

Conducting an assessment in the early stages has many benefits.

- The early assessment is the first of a *before* and *after* series. This helps you and your colleagues to see what progress is being achieved year after year.
- It provides much of the information required to set priorities and to create a sound initial plan—one that is based upon the organization's needs rather than a predetermined formula.
- It provides another opportunity to involve and to educate the top management team. Members need to learn about (and agree with) the assessment criteria beforehand, and this helps ensure an understanding of the full scope of a quality approach.
- If the assessment makes use of internal people to gather information, it offers an opportunity to involve and educate a cadre of middle managers. These individuals will be ideal members of the network that will be established later to provide ongoing support for the process. These people will

Exhibit 4.7 *Steps in conducting a thorough assessment process.*

1. Agree on the assessment strategy and criteria.
2. Educate management in the criteria (and adapt the criteria, if necessary).
3. Form an internal information-gathering team.
4. Train the team in the criteria and in the information gathering process.
5. Plan and conduct the information-gathering process.
6. Collate the information and write a factual report.
7. Have objective external experts validate and assess the facts and provide feedback to senior management.
8. Use this feedback as input to the strategic planning process, and plan the actions required to cause improvement.

quickly become committed to the changes inspired by their assessment work.

- An assessment should also help to identify and catalog existing initiatives that are under way. It will be vital to build upon these efforts, rather than to tear them down, and to involve the people who championed them.

How to conduct an assessment

There are many ways of organizing an assessment. These range all the way from a very quick and simple "what do we think of ourselves" to a rigorous and thorough examination of how the organization functions.

Exhibit 4.7 shows the typical sequence of events for a thorough assessment. This includes an external assessment portion, in which outside experts are used to validate the information gathered, to interpret this information, and to feed back the results to top management.

At the other extreme, a quick and inexpensive alternative is for the top management team to study and agree upon the assessment criteria and then conduct an initial broadbrush assessment under the guidance of an expert, using just the information in their heads. This can all be done in a day or so. This approach has the benefit that top management has full ownership of the findings and can begin to form an intuitive grasp of the power of these criteria.

However, this approach is superficial. It may, in a pinch, be used as the basis for action to address some of the most obvious issues, but it is best viewed merely as a precursor to a more thorough, factual, and objective assessment.

Tips on the assessment process

The assessment must be designed to provide well-informed and objective feedback on your situation and unbiased recommendations for action—not a disguised sales pitch. For this reason, it should probably not be conducted by a company that will be vying for other work in support of your improvement efforts. If, for example, you select the *Instantaneous Improvement International* consulting group to conduct the assessment, you may already be well on the way to committing your organization to their product, the *Instant Miracle Improvement System*®. This is not due only to commercial self-interest. When you have only a hammer, everything looks like a nail—you may not see the loose screws.

Rather than risk trying to create a force-fit with someone else's proprietary approach, you will do better to conduct a truly objective assessment, develop a plan based upon your organization's needs, and then select specific suppliers whose offerings fit your plan. (In chapter 8, "External Resources," we will discuss in more depth how to select and use outside resources.)

For the purposes of an initial assessment, you should consider

- Choosing nonproprietary assessment criteria, such as Baldrige or Canada Awards for Business Excellence (CABE) criteria. These are published as part of United States and Canadian national quality awards programs.*
- Using as assessors outside experts who have no vested interest in any particular implementation approach and who do not expect to become involved in implementation of their recommendations.

---

*Appendix D provides contact information to obtain details of Baldrige and other major quality awards.

- Deferring major decisions regarding other quality-related suppliers until it is clear what types of outside support you actually need.

Assessment criteria such as Baldrige are designed for the examination of high-performing companies with a mature quality process, and are, therefore, extremely thorough and detailed. For an initial assessment, the broad framework of these criteria will be valuable, but some of the detail will not be applicable. You may want to be selective in your use of the criteria, to focus on the most relevant parts, and to avoid confusing people with information that is not yet needed.

The initial assessment should also examine some issues that are so fundamental that they are not included in advanced criteria, such as the deployment of mission, vision, and values. Are these clear, consistent with a quality approach, well understood, and accepted throughout the organization? These basic questions should also be examined at the outset.

The assessment report

At the end of the assessment, management should have a report that includes

- A factual account (written by internal people) that sets out how the organization operates today.
- A list of existing strengths to build upon.
- A list of opportunities for improvement.
- A short list of recommendations regarding a course of action. Some of these recommendations will relate directly to the elements of a quality approach. For example, there might be no formal system for getting feedback from employees. Others will relate to side issues that could impact the implementation of a quality approach. For example, if a reorganization is pending, some of the recommendations might be related to accomplishing this in a manner that is consistent with the values, or reinforces a focus on the customer. The recommendations should flow logically from the observations, and should be presented as a basis for discussion and decision making by the management team, rather than as a prescription to be accepted blindly.

Reactions to the assessment findings
When the team members hear the assessment findings, they may be shocked—or they may show little surprise. They may say that the assessment merely confirmed their suspicions about many things. However, although this might be an honest reflection of what people feel, this apparent reaction understates the effect of the exercise. Even if the results evoke little surprise, the assessment may well have had a very significant impact on the team's thoughts and actions. Here's why.

As the team examines and accepts the criteria used for the assessment, people are already beginning to reexamine their views about what an ideal organization looks like; for example, what systems and what information they need to run the business properly, what key processes exist, and how these processes should be managed.

In addition, the information provided by the assessment may confirm shortcomings that were only suspected before, making it difficult now for management to deny the need for action. For example, having cheerfully agreed that employee morale is vital and should be examined as part of the assessment process, the management team members may discover that they are collectively doing a poor job in this area.

Perhaps the organization has great systems for sending messages to employees, but no way of discovering whether these messages were even received, let alone understood and accepted. Perhaps there is no system for gauging employee morale. Having set up the assessment process, the team cannot easily escape reality by ignoring or disagreeing with these findings.

Ensuring acceptance of the findings
By building a consensus on the vision—on what a great organization this *could* be—the team is already facing up to the current situation. However, there is always some risk that the management team will balk at facing reality. We all see someone younger and more attractive than ourselves when we look in the mirror. There is a risk that management may attack the assessment process, the data, the people—anything to avoid the disagreeable truth. There are several ways of minimizing this risk.

- The entire exercise should be designed to give the team a strong sense of ownership—of the process itself and of the outcomes. You should reinforce this ownership at every stage of the process by seeking managers' decision to conduct an assessment, by providing education on the criteria, by involving managers in the selection of the fact-finding team, and so on.

- You should ensure the credibility of the messenger(s) who finally deliver the findings. Some messages the team may find hard to accept from any source except highly-regarded outsiders.
- You should help the team set realistic expectations. Some of the following messages may help:

  "What we expect to get out of this is a lot of opportunities for improvement and some help in prioritizing our efforts."

  "This is a very stringent process designed for the scrutiny of world class organizations. Don't expect a great score!"

Moving forward

To know and not act is not to know.

—Wang Yang-min

The natural reaction to looking into this assessment "mirror" is a desire to take action, to seize opportunities, and to put right the most offensive blemishes. There is now an even stronger desire within the team to act. A proper, detailed plan is called for. This is progress!

There is also a deeper realization that this really is not a program. Issuing a policy statement and purchasing some motivational training will not do the trick. Management team members may realize, "Wow, this is much bigger than we imagined."

There also may be some dejection. "Somehow, this little snack we set out to eat has grown from a chicken to something more like a cow." At this point, no one will thank you for trying to explain that it's really an elephant. Better to focus attention again on the compelling reasons for change.

"Do we really want to do this?" the team may ask itself yet again. Don't rush things—it's the team's decision, not yours.

5

The Plan

Adventure is the result of poor planning.

—Colonel Blatchford Snell

CHAPTER CONTENTS

Possible elements of a plan for the transformation, factors in selection and sequencing, and how to review and validate the plan.

- Developing the plan
- Areas to consider in the plan
- A typical initial plan—sequence and rationale
- Verifying the plan
- Key success factors
- Designing the change process to endure in the long haul

Just as there is no recipe for securing top management commitment to the process, or for developing the plan, there is no standard road map for launching the change process. Every organization has to find its own way, geared to its own situation and its own needs.

However, there is no need to reinvent the wheel. You can learn much from looking at what others have done before. This will make it much easier for the senior management team to devise an approach that will work in your situation.

DEVELOPING THE PLAN

When the team has successfully digested the assessment results and built up the fortitude to press on, the stage is set for some thorough and detailed planning work. Here are some suggestions to bear in mind as you begin to devise the planning process that you will propose to your colleagues.

- You must involve the entire team sufficiently to secure its ownership of the end result. You may need to take the lead by developing a draft plan, but you must not aim for something so polished and complete that there is no room for your colleagues to contribute. They need to understand and agree with the logic, take part in decisions where choices exist, and work out the details in areas where they will be personally responsible.

  In order to ensure that others can contribute and take ownership, the draft plan should offer alternatives, and leave blanks where input from others is desirable. You might even include some controversial ideas that will ensure challenge and debate.

  Remember, that a plan which you may consider slightly flawed, but which your colleagues own is infinitely better than one which you consider perfect, but which they feel belongs to you. And, making mistakes is an integral part of learning. Who says that you're always right anyway? On the other hand, you must challenge anything that you feel is damaging or unworkable—if it is sufficiently important.
- Ownership lies not only in the development of the plan, but in the assignment of the tasks. Every one of the top management team members should end up with a well-defined piece of the action which they are personally responsible to implement. (This does not

imply that they are *solely* responsible. A team effort will usually be needed, often involving other senior managers. However, it is their job to lead this effort.)

• Link the improvement plan to the organization's objectives—the vision and the key priorities. This plan is only a means to an end. If the organization could get there by some other easier way, you would not be doing all this work. If the improvement plan will not help the organization reach these goals, it serves no purpose.

• Work toward integration of the improvement planning process and the overall strategic planning process into a single combined activity. If this is not possible at first (for example, for reasons of timing, or because the business planning process does not work well), at least ensure that the improvement plan gets built into the overall business plan and that improvement objectives are assigned in the same way as other objectives.

• Following this same thought, take full account of other current realities. The plan for the change process will consume significant time and effort, and some money. If you overload people or don't do your budget sums, the plan for change will probably be the loser during implementation. You cannot stop the business while you make the improvements. It is much better to thrash out the necessary compromises during the planning stage, rather than produce a plan that is not doable.

• Use the assessment findings. The assessment process is also a well from which management can draw ideas and set priorities for improvement year after year. The initial assessment, if properly done, will provide extensive input to the plan and greatly simplify the team's task.

• Build on existing initiatives. It is important to recognize the value of what has already been done, and to build on this. If you fail to do so, the new plan may be rejected as an emotional reflex by the very people who should be its strongest supporters—those who have already made attempts to create change for the better.

• Start small and simple, with actions you are confident of being able to execute.

The rest of this chapter steps through the various activities that could appear in the plan. Some of these activities may already be under way, and others may be a low priority for the moment. Our purpose is to consider why each is useful, how it may be tackled, and to gain some sense of sequence and timing.

AREAS TO CONSIDER IN THE PLAN

The table in Exhibit 5.1 shows many activities that typically appear in plans to implement a quality approach. There are many other activities that your organization could undertake that are not included in this list because they are too advanced, too difficult, or inappropriate for use in the early stages. However, the list in Exhibit 5.1 still contains far more activities than should appear in the initial plan. This is a list of what your team *may* do. We will now examine each of these and look at how to decide what you *will* do.

Customer relationships and focus

Customer relationships are usually the top priority activity in the plan because delivering value to customers is usually key to survival. Unfortunately, organizations tend to develop a self-centered mindset and so lose touch with their customers. Getting in touch with customers requires some new mechanisms as well as new attitudes. Over time, you may develop a wide range of methods for listening to customers, such as surveys of various types, focus groups, executive visits to customers, analysis of complaints and inquiries, and speaking to former customers who took their business elsewhere.

You will also need to develop ways of making sense of all of this customer input, and translating it into requirements for your products and services. In time, you may learn to use advanced methods of capturing and translating this information, such as quality function deployment (QFD). However, for your initial plan, you only need to decide which of the possible basic approaches will provide the most benefit, and focus your efforts on one or two such initiatives.

A formal survey of customers is often a good start. If this is done well, it will provide

- A high-level picture of what your customers like and don't like about doing business with you. Once the problem areas are identified, these can be investigated in more detail.
- Data regarding customers' priorities—what needs most to be improved from their perspective.
- Data regarding your customers' views of your competitors.

Exhibit 5.1 *Typical planning framework.*

Activity	Issues
Customer relationships and focus	• Who are our customers? • What do they need and want? • How are we doing in their eyes?
Employee relationships and involvement	• What are our employees' needs and concerns? • What do they need to do a great job? • How are we doing in their eyes?
Measurement and tracking	• What are the key outcomes that we need to quantify? • What intermediate outcomes will indicate progress? • How will this information be shared and used?
Awareness raising and education	• Where are we going? • How will we get there? • What can each individual do?
Process improvement projects	• Which processes offer the most leverage? • How do we organize and equip ourselves to tackle these?
Recognition and reward mechanisms	• Who is making a real effort? • How do we say thank you?
Supplier relationships	• Who are our key suppliers? • How can we make better use of their expertise? • How can we help them improve?
Organizational changes	• Does our structure support what we are trying to do?
Creation of a support infrastructure	• How do we ensure that everyone gets the help they need?
Education for the change agent(s)	• What may lie ahead? • What better methods exist?
Reassessment and replanning	• Where are we now? • How did we do? • What did we learn? • What should we do now?

This survey will also provide the basis for ongoing measurements of overall customer satisfaction—a valuable indicator of progress.

These initial steps are likely to raise customers' expectations, and it is critical that management respond to the information received. The customer survey plan does not end when the results are received. The appropriate responses must then be worked out.

Management's response to customers will be most effective if it is guided by logic and emotion (see Exhibit 5.2). This is an important consideration in deciding what methods to use.

For example, a thorough questionnaire survey might provide very reliable information about customer concerns and priorities. However, if managers see these survey results as just numbers on a page, and cannot relate them to real-life events, they may not pay much attention.

If, on the other hand, top managers go out directly to customers and gather some of the information themselves, the conversations they have may create more emotional energy—a deeper sense of concern about customer issues. There is nothing like a dressing-down by an important but irate customer to grab a complacent executive's attention. However, if managers only have such anecdotal information from a few customers, it may not be clear what the company should do to satisfy the important needs of most target customers. Statistically reliable data are also needed.

A good strategy for getting in touch with customers will operate on both fronts—it will reinforce top management's emotional commitment to customers (for example, by ensuring direct contact)

Exhibit 5.2 *The need for both logic and emotion.*

> Picture the challenge of launching a rocket to reach a distant planet. Huge amounts of energy are required, or the spacecraft will not even get off the ground. But an accurate guidance system is also needed, or it will go nowhere near its target. Both are equally essential.
>
> The entire improvement process is very much like this. It needs a combination of logic to provide direction, and emotion to generate energy.

and also provide management with statistically reliable information about what customers need and want.

There is another aspect of the customer relationship that may need urgent attention—the contractual side. Sometimes petty contractual conditions, designed to save your organization a few nickels or to avoid some perceived risk, can infuriate customers and damage the relationship out of all proportion to the sums of money involved. If you do have such problems, your initial customer survey work should quickly reveal this fact.

It is important to ensure that the "customer first" mind-set is reinforced at every opportunity—in the organization's mission, vision, and values; in the improvement plan and the tangible objectives chosen; in the setting of priorities for process improvement; in the measurement systems; and in the recognition and reward systems. Customer focus is not dealt with in just one section of the plan—it should be a recurring theme throughout.

Employee relationships and involvement

Effort put into developing knowledge and skills will be largely wasted if people feel alienated from the company or are prevented from contributing to the full extent of their capabilities. The plan may therefore include steps to get in touch with employees, and to ensure that they are given opportunities to contribute.

Getting in touch with employees is similar to getting in touch with customers. Some new mechanisms may be required, as well as new attitudes. There is a similar need to tap emotions (in order to create the energy required to take action) and to engage logic (to ensure that the actions are appropriate). For example, you might

- Institute a formal survey of employees to find out what their concerns are and the extent of their understanding of the organization's direction. This will also provide a measurement of morale at regular intervals—at least annually.
- Use team efforts, to solve problems or to improve processes, as a vehicle to involve people and to harness their ideas and creativity.

- Reexamine the pattern of face-to-face communications (like staff meetings) to help ensure that such encounters do take place and that communications are open and two-way.
- Reexamine complaints systems and other "safety valves" to ensure that employees have effective means of voicing concerns and seeking redress.
- Establish a program of "skip-level" meetings, in which people get to meet informally with a manager more senior than their own immediate boss and get to choose what they will discuss.

Management's response to the input obtained from such systems is crucial. Employees cannot be expected to care much about customers' problems if management clearly does not care about employees' concerns. For many frontline people, this new program will be considered to be just talk until management starts to correct the small, but long-standing, irritations that affect them.

Employees are also aware of many operating problems that management may have overlooked or chosen to ignore as trivial. Taken together, these may account for large amounts of chronic, costly waste which could be eliminated. However, the major cost of ignoring these is the loss of employees' involvement and pride in their work. Exhibit 5.3 illustrates the type of "minor" problem to which management must respond if it is to convince frontline employees that this new initiative is for real.

Union representatives can also play a positive role in helping management to keep in touch with the feelings and concerns of employees. Thus, union representatives must be made partners in the process.

Measurement

Like customer focus, measurement is an issue that needs to be tackled in many places within your plan. The key points are

- You absolutely need measurements as indicators of progress. You will never know for sure whether something is improving or not—whether your efforts are succeeding—unless you can measure something that will serve as an indicator.
- You need a *hierarchy* of measurements, ranging from high-level (for example, customer satisfaction and retention, market

Exhibit 5.3 *Fixing "minor" problems for employees.*

A company that employed many female operators in a low-wage industry was struggling to reduce error rates. Having had no success with motivational programs, the company asked the operators, via a questionnaire, what work-related problems caused errors. Managers were mystified to find the response on many forms, "No space for purses." Managers had to go and ask the operators what was meant.

Management learned that, in order to avoid theft, the female operators had to keep glancing down at their purses, which were kept on the floor by their feet. This was very distracting, yet these jobs required sustained concentration. The solution was to provide a storage bin for personal possessions in each work station. Error rates dropped immediately.

In fact, the employees had been asking for a solution to this problem for many years, but management had always ignored such "whining" about "trivial matters." When management listened this time and responded, this signaled clearly to employees that something had changed.

share, employee morale, total labor hours per unit) to detailed (for example, invoicing error rate, department B equipment downtime).

- The higher-level measurements should *relate directly to the organization's goals.* Some may relate to important operating parameters that you have to hold stable. Others will relate to the vital few areas that you want to improve, because these are the keys to success. In other words, many key measurements should come right out of the strategic plan for the organization (which includes the improvement plan). You need to select quantifiable objectives during the strategic planning process.
- This hierarchy of measurements needs to be *linked,* so that the detailed, lower-level measurements are related to the higher-level ones, and movement of lower-level indicators predicts movement of the higher-level ones. For example, you should have measurements for each of the few key issues that drive customer satisfaction. When the process improvement

teams succeed in moving these indicators in the desired direction, you can be sure that all of their efforts are contributing to improved overall customer satisfaction.

- The measurement system hierarchy is *constructed in a certain direction*—from the high-level measurements representing the goals, down to the detailed ones. The customer-related measurement system is constructed looking into the organization from the outside—working from the customer's perspective to the perspective of an individual supervisor or work group.
- The measurements are almost all *related to systems and processes.* The high-level measurements may be thought of as indicators of the performance of the organization as a system. Many of these high-level measurements relate directly to key process, while the lower-level indicators mostly relate directly to smaller subprocesses.
- *Every measurement should be someone's responsibility*—both to report on and to hold stable or move in the desired direction. The senior management team members may simply divide up the high-level measurements between them, so that each person will have some included in their performance plan. Responsibility for lower-level measurements will be assigned to people from other levels in the organization. This will naturally result from cascading of the high-level objectives and from the process ownership responsibilities assigned.
- Measurements should be *plotted and displayed graphically,* to enable trends to be separated from isolated incidents or random changes.
- Every measurement should be *useful as a basis for action.* If variations do not point to the need for action, the measurement is of little value.
- *Who receives the information, and when, is as important as what is measured.* There is little point in having the finest measurement system in the world if the people who are best placed to act on the results aren't getting the information, or receive it too late.

Building a hierarchy of logically linked and relevant measurements is a very big task. So you may have to do this progressively, starting with the measurements associated with the highest-priority organizational goals—like customer satisfaction.

These measurements—and the objectives that people have to keep them stable or make them move in the right direction—create a desire for ways to improve. People start asking questions like "How can we reduce these error rates?" The desire for knowledge and skills grows. Education is sought as a means to achieving the goals. These two must go in parallel—you should not hold people accountable for things they don't know how to do.

Awareness raising and education

There are typically two types of awareness raising and education efforts. Some are designed to reach all employees, and others are more targeted.

Universal awareness raising and education

Some awareness raising and basic education is generally required for all employees. This typically sets out to accomplish some of the following:

- To communicate and reinforce the vision and mission, and to explain why change is necessary
- To reinforce some underlying values or principles, such as the need for mutual respect and cooperation within the organization (and to explain how the people in charge are holding themselves accountable for their adherence to these values)
- To explain quality improvement principles
- To develop the interpersonal skills required to work in a fashion consistent with the values, and to ensure successful meetings and team efforts
- To provide the tools and techniques required for basic quality improvement techniques, such as data-driven problem solving and process improvement

A variety of delivery methods may be used, ranging from formal classroom sessions to the distribution of books, articles, and videos. Direct sharing and discussion of information on the job is one of the most effective methods and the manager's role in any kind of training is critical. Even lengthy or highly technical training, which may not be delivered by people's own managers, is of little value unless the manager understands the content and actively supports use of the knowledge and skills in the workplace.

The content and the method of delivery should be adjusted to the needs of the target audience. For example, management may require detailed explanations of the company goals and strategies, while frontline employees may prefer a summary of this type of strategic information, but more information on tools and techniques.

The execution of this awareness raising and education will be progressive, and it may take some time. Remember, to frontline people, their supervisor *is* the organization. If the company newsletter says "Give us your ideas," but the supervisor says "You're not paid to think," then nothing has changed in their lives.

For this reason, when specific behavior changes are being sought, it is often best to work down through the ranks progressively, checking at each level that the message is being received and understood and that the training is being put into practice, before going to the next level.

Universal education and awareness raising are essential parts of the overall plan. Some information must be shared with all employees. Providing basic skills training for all employees (for example, in problem-solving methods and tools) should be a goal—at least in the medium term. Those who do not receive such training are being shut out—denied the opportunity to make a contribution. How can they help improve the workplace if they have no method for doing so?

Targeted educational efforts

In order to secure early tangible results with the least investment or delay, the initial education efforts need to be targeted. People who are going to play a leading role in the process, such as members of problem-solving and process-improvement teams, require early intensive training in the techniques that they will use. They may also be introduced to more advanced techniques that others don't need. This type of training is discussed in the next section on process-improvement efforts. A typical and effective combination of early targeted training is

- A combination of hard and soft training (for example, interpersonal/team skills combined with problem solving or process improvement),

- Provided to groups of people who will use it immediately (for example, problem-solving teams),
- In situations where some tangible objective has been set (for example, to eliminate a specific problem, or to eliminate some specific type of waste).

As skills and knowledge come under closer scrutiny, there is often a realization that certain groups of employees need to develop specific job-related knowledge and skills. For example, salespeople may need more product knowledge, operators may need to hone their skills in equipment maintenance, finance people may need to learn about activity-based costing, managers may need to develop better planning skills.

Sometimes, there is also a need to remedy shortcomings in numeracy and literacy, so that all employees can read and understand procedures, work with numerical data, and take part in problem solving. If this is the case, then it may be better to carry out such remedial work before investing in problem solving and other training that may simply be lost without these basic skills. How will you know that such training is required? Formal needs analysis is one way. However, if given half a chance, employees will often be able to tell you themselves. Involve them in the decisions regarding what training they need.

Generally speaking, senior management people need the same education as is given to everyone else, and they need to receive it first. They also have the same need as anyone else for relevance—the knowledge and skills they acquire must help them to perform their work. They also need learning methods that make the best possible use of their time. Exhibit 5.4 describes the initial training cascade conducted by Xerox in the mid-80s. This is still considered a model of how to carry out this initial training thoroughly and effectively, and in a way that links training to real-life issues in the workplace.

Process improvement projects

Process improvement is the engine that generates tangible gains in quality and productivity and leads to improved products and service to customers. In mature quality organizations, process thinking and process-improvement methods permeate the entire culture and are reflected in the structure and the mode of operation. In these companies, managers are expected to have the skills required to orchestrate process-improvement work. So, demonstrated proficiency in these skills is naturally a prerequisite for becoming a manager. Exhibit 5.5 gives an example of process management in action, and 5.6 illustrates one of the essential tools—flowcharting.

However, for an organization just beginning to learn about quality, few people at any level understand even the basic concepts

Exhibit 5.4 *The Xerox training cascade.*

When Xerox's *Leadership Through Quality* initiative was launched in 1984, a cascade process was used to train everyone in the company, from top management to frontline people. A five-day class (with a day of prework) was designed, which had participants work on a real-life issue from their own department, using quality methods and tools. The training was given to family groups—that is, groups of people who normally work together, and their manager or supervisor. The content included the need for change and Xerox's plans, basic quality principles, problem-solving methodology and tools, and a process-improvement methodology.

The class was led by a facilitator, but the manager took part in all of the training and led some segments. The first work group trained was the president of Xerox Corporation and his direct reports. Each of these managers then took part in the training of their own direct reports, and so on down through the organization.

In this way, each manager learned twice—once with his or her peers as a class participant, and once as a manager, teaching his or her people. To ensure that the training was effective, each manager would check that the skills learned were being applied. This approach was known as *LUTI,* or *Learn-Use-Teach-Inspect.*

Xerox went on to institutionalize the content of this training by making the mastery of these skills a part of individual performance appraisals and a requirement for promotion.

of process management. Therefore, you must find ways to build progressively from this starting point.

Aiming for early wins

A typical strategy is to set up a few initial cross-functional teams, with the aim of securing early benefits and developing some experience and expertise in process improvement or problem-solving methods. The teams work best when they have a senior management sponsor to provide support and relevance, and a facilitator to keep the teams on track.

These teams are typically among the first to receive training in quality improvement tools and techniques, and they typically deliver the first tangible gains. These early wins help to sustain the mandate for the process, and provide some insurance against setbacks in other areas. The teams also act as a training ground for people who will become key players in supporting further efforts.

The first teams usually tackle systemic issues, which are critical to survival (for example, because they are important to customers) or have a high payback. These issues are typically chronic process problems that have proven, in the past, to be unsolvable using traditional management methods. Exhibit 5.7 offers criteria for selecting early improvement projects—whether problem solving (usually within a process) or process improvement.

A process that cycles slowly—such as designing a complete new product—takes much longer to demonstrate tangible results. If you need to work within such a setting, you might look for some troublesome subprocesses that are frequently repeated, such as reporting and correcting design problems or the design of subcomponents.

An ambitious and comprehensive approach

A more ambitious approach is the cascade. As previously described in Exhibit 5.4, this is an example of combining training with the formation of teams (in family groups). It also ensures that the training is used immediately (starting in the classroom) to tackle real-life issues. However, not every organization has the discipline to execute such a thorough and comprehensive deployment, or is willing to commit the resources required.

Exhibit 5.5 *Process management in action.*

In its efforts to win back market share, Ideal Company discovers that customers find its invoices inaccurate and difficult to understand. Some customers waste so much time querying invoices that they are now seeking alternative suppliers. A solution *must* be found.

What should top management do? Give a pep talk to people in the billing department? Offer bonuses for more accurate work? Replace John Bills, the manager of the billing department? Buy a bigger computer? Ideal knows that none of these actions is a solution. This company recognizes that billing is a business *process,* and it is the process that is malfunctioning, not the people, nor the manager.

So ideal sets up a cross-functional team of people from all the different areas involved in the billing process, such as order entry and shipping, as well as the billing department itself. The team members have already learned a formal method of studying and improving a process, and they go straight into action. Assured by their managers that they are allotted the time to do this, they are enthusiastic about the task because the current process causes them endless hassle and wasted effort.

First, the team does some research. They gather information from customers about exactly what is wrong with the invoices, classify these problems, and figure out which ones upset customers the most. They also draw flowcharts of the process, to create a picture of how it is supposed to work today —including the steps that involve order entry and shipping. They find that everyone has a different idea of what should happen, and there are many built-in sources of error.

The team can already see many problems in the process, and lots of room for improvement. Armed with this information, team members select the top priority problem to fix—pricing accuracy. They brainstorm possible causes, and begin to investigate the most likely root causes. The main root causes prove to be that the order entry and sales people often have different versions of the product catalog, and that the

Exhibit 5.5 *Process management in action (continued).*

system for communicating price changes doesn't reach everyone who needs to know.

These problems are solved by a software change, some training, and a new pricing-change procedure. The team's monitoring measurements show a decrease of 40 percent in invoice errors over the following three weeks. After six months, other actions by the team have eliminated pricing errors and two other main causes of customer dissatisfaction.

John Bills, who was assigned the task of process manager, will track the billing accuracy measurement in the future, ensure that the problems do not recur, and reconvene the team if required. This success is publicized, and team members are asked to explain their methods to new hires, customers and visitors. And the president of Ideal gives them a pep talk to thank them for their outstanding team effort.

A typical approach

A third approach is to use a combination of family-group problem-solving teams involving frontline people, and cross-functional teams with more management participation. The cross-functional teams may use process improvement to tackle business processes (like order fulfillment), or problem solving to tackle some well-defined, but major, problem (like inaccurate pricing on invoices). Or, the team may simply take on some relevant project (like establishing a recognition system for employees).

A mix of approaches like this can yield some quick small gains, while still working toward bigger, longer-term big gains. This type of mix also can provide opportunities for involvement and learning for people at all levels in the organization.

The diagram in Exhibit 5.10 shows how these management-owned big efforts relate to the many small efforts that are owned more by frontline people, or that may even be spontaneous. Both types of effort will be more productive if some clear goals exist for the organization.

The large efforts are personally directed by management, to be aligned with these goals. In the early stages, the easiest way of aligning

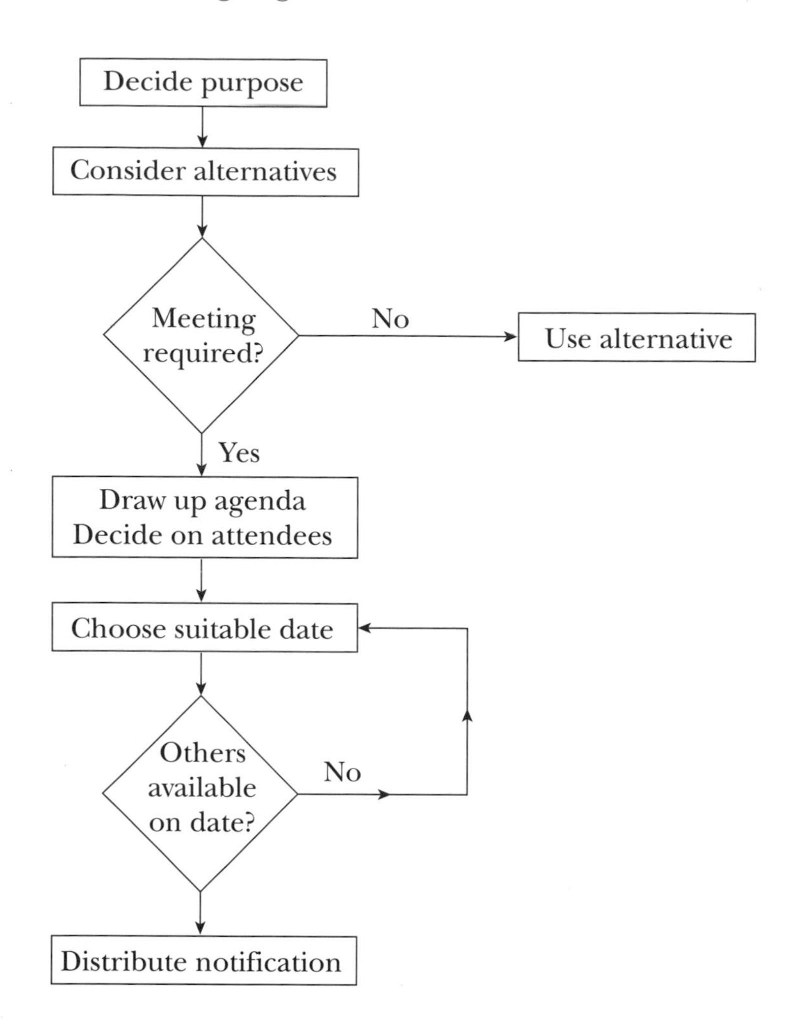

Exhibit 5.6 *A simple process flowchart: Planning a meeting.*

the many small efforts is to ensure that the goals are communicated and that everyone understands and accepts them. In this way, the local and individual efforts of frontline people will also support what management is trying to achieve. Without this communication, when employees begin to get involved and tackle problems, the result may be a hodgepodge of activities with widely divergent, or even conflicting, objectives. For example, employees may start work on improving some activity that, unknown to them, is to be eliminated because the customers do not want it.

Exhibit 5.7 *Criteria for selecting early improvement projects.*

- It is not too ambitious (not *world hunger*).
- There is a real desire to tackle this issue (and, hence, energy and commitment to do some work).
- There is potential for real, tangible benefits (and, hence, early wins)
- The process repeats frequently (preferably).
- The process is established (not brand-new or just being created).

Exhibit 5.8 *Types of teams and methodologies.*

Membership	• Family group • Cross-functional
Methodology	• Problem solving • Process improvement • Project

The communication of these company goals generally takes place through the normal channels, such as meetings and newsletters. However, the goals may also be communicated and shared in a more systematic way through the system for objective setting and performance appraisal.

Process management is a substantial body of knowledge that lies at the heart of quality improvement. This chapter presents little more than a basic outline. This subject is an area for intensive study, to make sure that you know what you are doing.

Recognition and reward mechanisms

The most important forms of reward are implicit in the task; for example, like producing work you can feel proud of, learning new skills, or seeing your ideas being put into practice. Deming refers to

Exhibit 5.9 *Problem solving and process improvement contrasted.*

Problem solving	Process improvement
• Reactive—start with an identified or suspected problem	• Proactive—start with a process selected for improvement
• Simple methodology, limited training needs	• More complex methodology—builds upon problem-solving skills
• Gets results quickly, can accumulate many small gains	• Takes longer to complete an improvement cycle and get results, but larger gains are possible
• Risk of fragmentation (through many separate, unconnected efforts) • Risk of suboptimization (improving part of a process while unintentionally making some other part worse)	• Integrative approach—pulls together and streamlines activities that are fragmented
• Can be done quickly and cheaply; hence, may be initiated and owned by frontline people	• Needs significant time and resources; hence, more likely to be initiated, directed, and, owned by management

such intrinsic rewards as *the joy of work* and *the joy of learning,* and these are powerful motivators of human behavior.

Recognition is another important motivator. The most important form of recognition is to be respected as an individual—to be listened to and to be heard. This reinforces our feelings of self-worth and self-respect. Being appreciated and thanked for our ideas and for our efforts is a natural next step.

These are powerful mechanisms, but they are often nullified by other powerful negative mechanisms operating in the workplace. For example, there are barriers to performing good work that are highly demotivating, and there are punishments for helpful behavior like highlighting problems and volunteering ideas.

AFTER 12–18 MONTHS

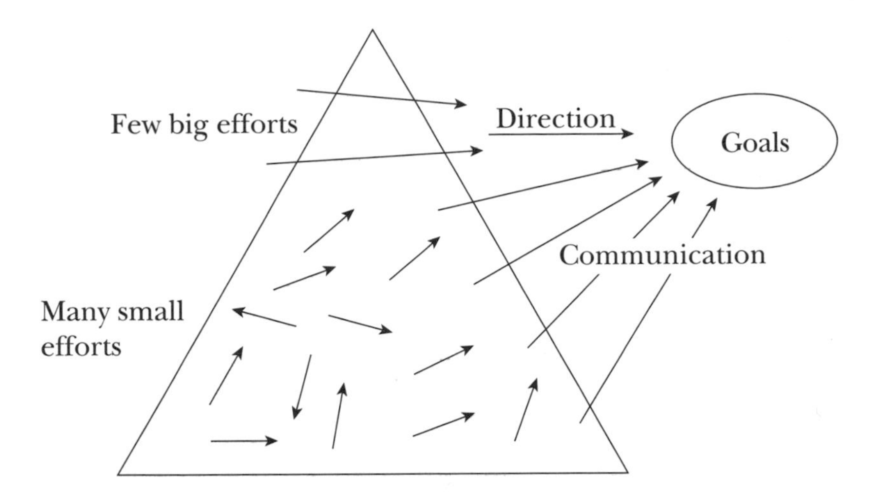

Exhibit 5.10 *Improvement activities linked by communications.*

AFTER 2–3 YEARS

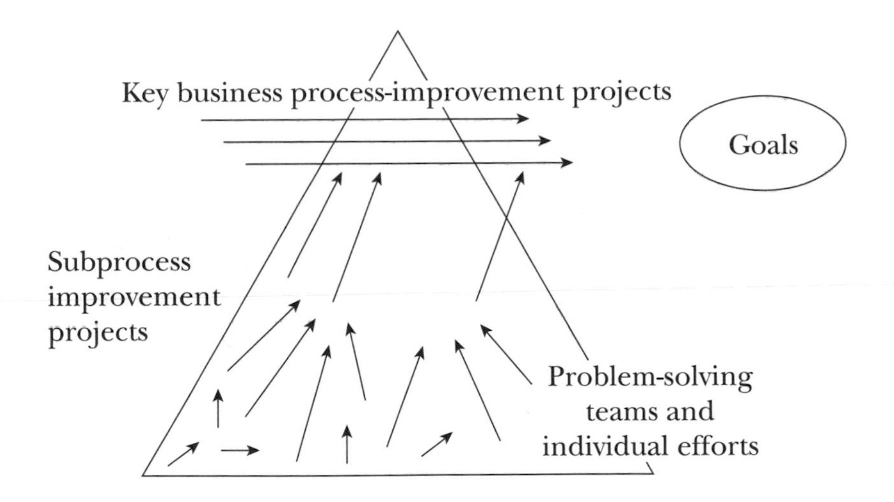

Exhibit 5.11 *Improvement activities linked by process management.*

So, you need to employ many ways of supporting the desired changes in behavior. For example, you may

- Remove the barriers that prevent people from taking a natural pleasure in performing good work.
- Build an atmosphere of mutual respect in which people value themselves and their work.
- Provide explicit recognition and reward systems in which there are clear definitions of what will be recognized and rewarded.

Your first priority is to ensure that the efforts and successes of the early enthusiasts are recognized. Thank those who put in the effort on teams and help publicize their methods for others to emulate. This can be achieved without extensive work on formal systems; for example, by building recognition into the mandate of everyone who is supporting the process, including team sponsors and facilitators. It is their job to anticipate recognition opportunities and to quietly arrange something appropriate.

It is also essential to reexamine the formal systems for performance appraisal and for promotion, to ensure that these are consistent with the declared values and that they reinforce the desired behaviors. Usually major changes are needed in these systems.

Typical aims are to ensure that

- Team efforts are rewarded—rather than the efforts of individuals who strive to make themselves look good, perhaps at the expense of others.
- Favorable appraisals and promotions are earned by providing service to customers and maintaining employee morale— not from pandering to the whims of one's manager.
- Pay raises and career opportunities result from working to achieve more with fewer people and less money—not from empire-building to secure an ever-larger budget and headcount. Exhibit 5.12 gives an example of how reward systems may work against the best interests of the organization.

These aims can be achieved by following some guiding principles in the design of all reward and recognition systems; for example,

- An important aim is to make each individual feel valued and respected as a person.
- There should be goals that everyone shares.
- Everyone involved in a team effort should receive the same recognition or bonus for this effort.
- There should be recognition for appropriate effort and behavior—such as, use of quality improvement tools or completing a phase of an improvement project—as well as for successful outcomes.
- There must be support systems, encouragement, and recognition for the recognizers. It takes some effort to look for actions worthy of recognition and to take the appropriate action, however small. Why should busy managers and supervisors bother to do this, if they never receive any positive feedback for these efforts?

It is a common mistake to naively place too much faith in monetary incentives and overlook nonmonetary rewards like simple appreciation. Both reward and recognition can be used to influence behavior, but there is only so much money available to give out.

In contrast, there is no limit to the amount of appreciation that management can generate—if it is understood that this is an important part of a manager's job. Most people value appreciation a great deal when it is sincere and comes from individuals whom they respect. Showing employees the appreciation they deserve can generate energy like throwing gasoline on a fire—and it is free!

Recognition can be institutionalized to a degree with formal programs, but it becomes ineffective or even counterproductive unless it remains prompt, personal, and sincere. Small informal gestures of appreciation should become part of the daily routine, and these are just as important as the formal programs.

However, showing appreciation is a skill that managers often have not been encouraged to acquire. Learning to say "thank you" can be almost as hard as apologizing when you have been taught not to show doubt, weakness, or emotion to others—especially to those who you think should look up to you.

Your organization can develop a more appreciative style of working. But, like all other aspects of a cultural change, this takes time and energy, and it has to start at the top. Fortunately, most of

Exhibit 5.12 *Negative rewards by job evaluation.*

The new quality manager inherited a department that she discovered was at war with almost everyone—certainly with production, procurement, and design. An army of inspectors and auditors scrutinized and double-checked other people's work, and justified their existence by the large number of errors, which they found month after month. The quality department often stopped production and halted shipments, causing havoc until its demands were met. The cost was enormous, yet product quality had not improved at all in the past decade.

Over the next couple of years, the new manager put an end to most of this. She helped other departments learn how to manage the quality of their own work by providing training for their supervisors and operators and by focusing on preventive methods. She reassigned most of her inspectors to the line departments, where they retrained and took on other more productive work. By the end of her second year, she had a smaller department, conducting training and process audits. Product quality was improved by an order of magnitude, and reductions in scrap and rework had saved the company millions of dollars. The line was rarely stopped for long now, and it was the production people themselves who stopped it when necessary.

At this point, the quality manager's position was re-evaluated using the company's formal job evaluation program. In this scheme, each position in the organization was ranked according to a formula in which size of budget and levels of authority figured strongly. The previous incumbent had done well from this system, based upon a large department and the frequently used authority to stop all operations.

The job-evaluation formula resulted in the manager's position—and those who reported to her—being down-graded and their compensation frozen. So she left, and went on to employ her talents successfully elsewhere.

the top management team members will love handing out sincere praise and appreciation to people who are improving the organization—if you help them remember to do so. Saying "thank you" to each other isn't so bad either, once you get used to it.

Supplier relationships

Procurement and the management of supplier quality is an entire field of study in its own right. The following are typical strategies for working with suppliers, which are generally pursued over a time frame of many years.

- Changing purchasing criteria to take account of life cycle costs rather than just purchasing price. This reveals how purchasing poor-quality equipment and materials is wasteful and costly, even at very low purchase prices.
- Identifying critical categories of purchases and choosing one or two highly competent suppliers for each of these.
- Reducing the total number of suppliers in order to develop closer working relationships with a critical few, and to help these suppliers to improve quality and productivity and reduce costs.
- Working toward a cooperative long-term relationship with key suppliers, perhaps reinforced by different contractual arrangements that give them more stability and more incentive to improve and invest in the future.
- Involving key suppliers in planning and design work in order to make best use of their expertise and capabilities.

The importance of suppliers varies enormously from one organization to another. For an aircraft or automobile manufacturer, suppliers are the lifeblood of the business—these companies cannot be better than their suppliers. On the other hand, a legal firm or an advertising agency may not purchase anything more critical than space and office equipment. These may be important purchases, but they are probably not the key to success in this type of business.

Many organizations do not include suppliers in their initial improvement plans because their suppliers are noncritical, or they feel the need to set their own house in order first. However, if suppliers are crucial to your organization, then some consideration should be given to them from the outset.

It is usually wise to start by obtaining your suppliers' view of the world, rather than assuming that your organization is perfect and that all your problems with suppliers' goods and services must be their fault.

If you are fortunate, some of your existing key suppliers may be more advanced on the quality journey. Then they can become valuable sources of expertise to help your organization improve.

Organizational changes

Once you have made some progress and begun to get a better feel for systems and processes, it may make sense to reexamine the shape of the organization itself. A structure that makes great sense when pursuing economies of scale and close control may suddenly look unwieldy if the aim is to achieve flexibility and responsiveness to customers.

So, organizational changes may also be required. Typically these are designed to create a shift away from specialized functional departments working independently, to cross-functional team efforts working within a process that is focused on specific customer needs. Such changes not only can be highly effective, but demonstrate that management is serious about improving customer service and making key processes work.

Creating a support infrastructure

What do we mean by a *support infrastructure?* This is the framework of responsibilities and resources required within the organization to sustain the improvement efforts. The support infrastructure has two complementary dimensions (see Exhibit 5.13).

1. The hierarchy of line management that drives the process by cascading objectives, regularly reviewing performance against these, and requiring evidence of effort and progress

2. The network of staff people who assist the process by showing people how these objectives can be accomplished, for example, by the use of appropriate methods, tools, and techniques

The management hierarchy includes

- The top management team, operating as the planning and steering group for the entire process. We have dubbed this team the Quality Council. It is in this forum that senior management develops and approves the plan, reviews progress on a regular basis, and recognizes outstanding efforts.

  To ensure full deployment and follow-through, similar planning and review meetings are required at other levels in the organization, either as an integral part of the existing management system, or (in the short term) as an additional parallel system.

The support network includes

- The officially appointed Change Agent.
- Depending upon the size and shape of the organization, a few or many other people may be involved in supporting roles, full time or part time.

In a large organization with independent or semiautonomous divisions, each may have its own divisional Quality Council and Change Agent.

The network exists only to support the implementation of the plan, so the scope and size of the network should be determined by the requirements of the plan. Quality is all about doing more with less and eliminating waste—so you should seek for this network only the resources that are essential. Besides being wasteful and setting a bad example, having too many support resources can create as many problems as having too few to support the implementation of the plan.

In chapter 7 we will discuss the support network in greater detail—how to establish it, and how to nurture and develop it.

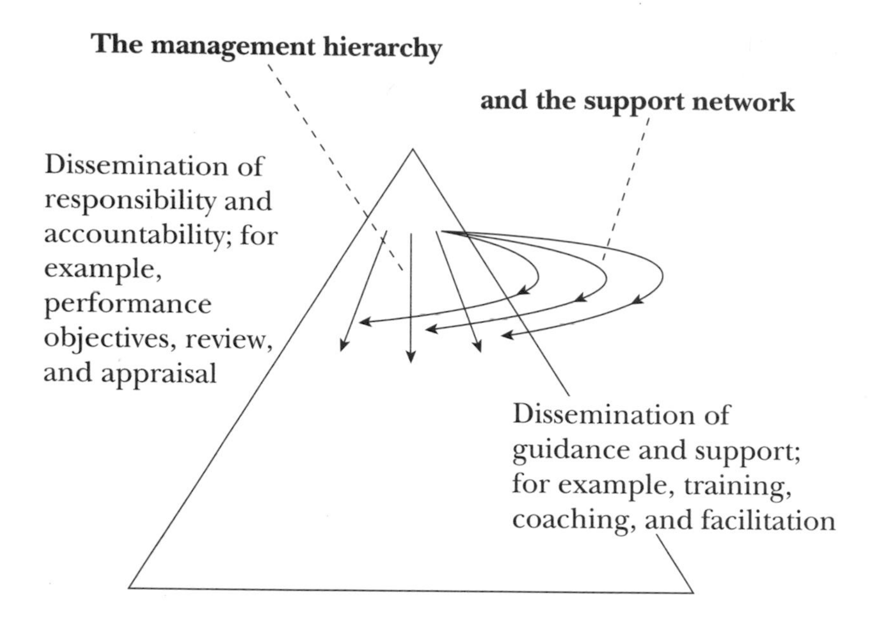

Exhibit 5.13 *The management hierarchy and the support network.*

Education for the change agents

This activity is listed here, not because it is difficult, complicated, or costly, but as a placeholder to make sure it is not forgotten. It is a reminder to plan for the education and development of the official Change Agent and all the other change agents in support roles.

The ongoing personal education of the change agents should be planned and budgeted, and some kind of timetable for this set out. It is too easy for these individuals to become so involved (and so hard-pressed) that personal development is forgotten. But it is vital for these individuals to stay on a steep learning curve—in order to give the organization the sound guidance it needs on increasingly advanced issues, and to avoid the difficulties that can be caused by poor decisions.

The president also needs to recognize how important this is. The army doesn't let scouts get bogged down with the main forces—it frees them to move ahead and survey the lay of the land.

An education plan will also help the Change Agent to build an external network of contacts, both for sharing of expertise and for personal support. These networks are described further in chapters 8 and 9.

Reassessment and replanning

The assessment and planning activities described at the start of this chapter are not just required to produce your first plan. They are required to produce the next version of this plan and every version after it. To ensure that this happens, *future assessments should be included in your plan as an integral part of the planning cycle.*

You may not need or want a comprehensive, organization-wide assessment every year, but you do want to ensure that the management team "looks in the mirror" at regular intervals. So you should plan for some kind of broad reexamination of the change process during every planning cycle.

FINALIZING AND REVIEWING THE PLAN

A journey of a thousand miles must begin with a single step.

—Lao-tzu

Selecting the areas to start on

Perhaps your head is spinning at the number of things to consider for the initial plan. "This isn't an elephant," you say, "it's a whale!" Don't be dismayed. You are going to start small and simple, with manageable tasks that you can achieve. These will provide learning opportunities and some early successes, and will build the confidence to tackle more ambitious tasks. It doesn't matter how long the journey, you only need to be prepared to take the very next step ahead of you.

Once the top management team understands the main options, you can begin to narrow these down. This will not be as difficult then as it might seem now. Here are some reasons why the task of creating the plan may not be as daunting as it appears.

• The initial assessment will provide a wealth of information and some clear direction.

• You will work to make this a team effort. How can you possibly figure out on your own what to do in all of these areas and what the priorities should be? You cannot. You have been hired as the expert? Yes, but you are not all-seeing and all-knowing.

If the team is demanding a plan by yesterday so that it can get started, try to explain that it is the team's plan, not yours, and that there is too much involved for you to figure it out all on your own. Agree on a process to create the plan.

You may agree on some categorization of the issues and parceling out of responsibilities for these. Reward and recognition? Sounds like a good choice for the human resources person. Customer focus? A perfect choice for the vice president of marketing. If it achieves nothing else, this approach will slow things down enough to allow for some learning, and to avoid firing before the target has been sighted.

Your role will then be a supporting and integrating one, working with each individual to help him or her stay on track for his or her portion of the plan. You will also work with the team as a whole to make sure that the overall plan makes sense and that the various elements are linked together.

• The initial assessment will help you to determine initial priorities, based upon the current status of the organization.

• When you and your colleagues have explored the main options, considered the cost and effort involved in each promising activity, and decided which must follow others in some kind of sequence, then you will be in an even better position to prioritize. Exhibit 5.14 offers some ideas regarding priorities and timing.

• Your initial plan is just that—a first attempt at organized, systematic improvement. No one ever got started on this type of major change without some stumbles and setbacks. If you are clear about what each part of your plan is intended to accomplish, however, you can monitor the effects closely, and quickly identify when something is not working. You should expect to have to make some mid-course corrections.

• Your plan should address only those high-priority issues that you believe you can tackle with the time and resources available.

All the other improvement activities that you would like to do at some time must be put aside and ignored for the moment, to enable efforts to be focused on the vital few activities identified in the plan.

Exhibit 5.15 gives an example of a simple launch plan. This plan might be perfect for someone, but it's unlikely that yours will be exactly the same. For example, this plan is based upon a belief that employee relations are good—or at least good enough for the initial attention to be focused on the external customer. If this belief is incorrect, then little progress will be made, because employees will be unwilling or unable to respond to the changes required to meet customer needs.

Here's the logic that might lie behind this simple plan. The assessment results indicate that in this organization, at this time,

- The most urgent need is to get in touch with customers (and the sales results underscore this).
- Some of the key operating processes are poorly understood (and these are probably major sources of customer-affecting problems).
- The organization has no reservoir of expertise in systematic problem solving or process improvement.

The idea, therefore, is to proceed as follows:

- Immediately start to develop some expertise by training a few teams to work on obvious chronic problems (and selecting customer-affecting problems).
- In parallel, start work to contact customers and to get better feedback from them—to rebuild relationships, to determine their needs, expectations and priorities, and to get their views of the service they are receiving.
- When the customer data have been analyzed, identify the high-leverage customer-affecting issues and the related processes, and start the teams working on these.

Exhibit 5.14 *Priority and timing considerations in the plan.*

Reestablishing the mission, vision, and values	Usually need to tackle this first with the top management team, then progressively widen the circle.
Customer relationships and focus	A top priority is to identify key customer issues—these will drive the improvement efforts.
Employee relationships and involvement	A top priority is to identify the key barriers preventing people from doing their jobs right and reasons why they feel unimportant or alienated.
Measurement and information	Initial efforts should be linked to key customer needs and improvement objectives. Overhaul of the complete system will take some time.
Process-improvement and problem-solving projects	A top priority. Start small with a few carefully chosen efforts. This work is essential to achieve tangible gains and early wins, and to learn the approach.
Creating a support infrastructure	Gear the timing and scope of this to the needs of the plan.
Awareness raising and education	Focus initially on management and on support of early team-based efforts (process improvement and problem solving). Use top-down flow to reach the whole organization eventually.
Recognition and reward mechanisms	The priority is prompt, nonmonetary recognition of early team-based efforts. Recognition and reward can be dysfunctional if offered prior to education and understanding.
Supplier relationships	Usually tackled later. Priority and timing depend upon how critical suppliers are to the organization.
Organizational changes	Address once the process structure of the organization is better understood.
Education for the change agent(s)	Ongoing—must not be overlooked.
Reassessment and replanning	Essential to plan for this. Timing is determined by need to integrate into normal planning cycles. There may be a need to rationalize the current business planning process.

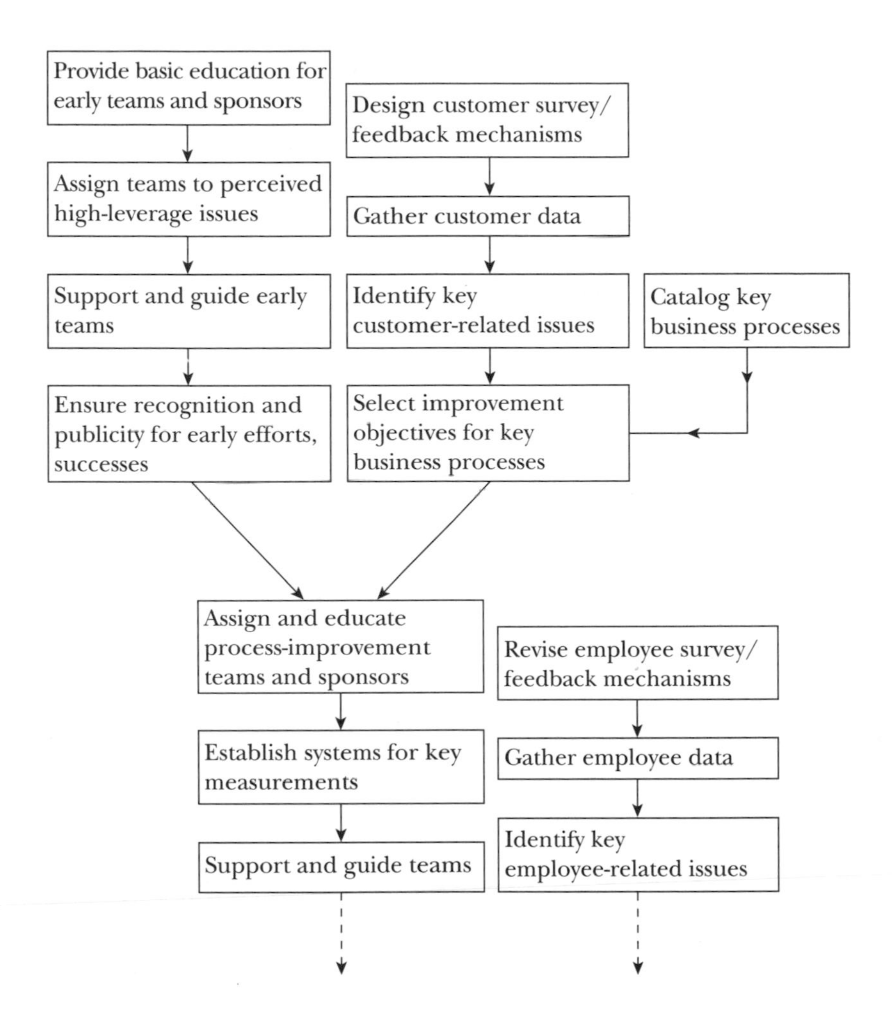

Exhibit 5.15 *Some elements of a simple launch plan.*

Some of the other desirable dimensions of the change process are incorporated to some degree.

- Each of the top management team members is involved and has some responsibilities; for example, for designing the customer-feedback processes, for visiting customers, for selecting the teams, for organizing the training, for sponsoring or taking part in team efforts, and so on.
- The methodologies taught to the teams ensure that measurements are put in place to allow progress to be tracked and improvements subsequently monitored for permanence.
- There is wide communication, to all employees, of the assessment results, of the initial plan and the underlying logic, and of the progress of the teams.
- Responsibility has been assigned to team sponsors for making sure that the people who participate are appropriately recognized.
- There are postmortems and feedback mechanisms built into all of the activities to ensure that these questions are being asked regularly: "How is this working?," "What have we learned?" and "What could we do better?"
- The Quality Council has regular meetings planned to review overall progress, to review and recognize team efforts, to examine various indicators of how well its approach is working, and to work on a long-term plan.

Being prepared for rainy days

With the careful preparation that has been done, with some key sponsors committed and signed up, with the top management team beginning to unite behind a well thought-out plan—you can be confident of some successes. However, you cannot afford to become complacent, and you should be mentally prepared for setbacks, too. Internal and external events will almost certainly conspire to work against your efforts at times.

This is like rainy weather—you may not like it, but you don't feel astonished or betrayed when it comes along, and you are prepared for it, so your life isn't disrupted. To illustrate, Exhibit 5.16 provides some examples of typical events.

None of these is an everyday event, but none is very unusual either. If you add up how frequently each of these might be expected to occur during, say, a five-year period, a sobering fact emerges: *During any given 12-month period, it is likely that some kind of significant event will occur that has an impact on the change process.* This means that you have to expect and plan for these events just like you would plan for storms when designing a building—even though you may not know exactly when such conditions will occur.

Fortunately, there are quite a few things you can and should do, right from the start. Most of these strategies can be implemented without significant changes or additions to the plan. In most cases, all that is required is to tackle an already planned activity in a certain way.

It's also worth noting that some unexpected events will occur that will help, or at least can be turned to good use. For example, a crisis may be used to dispel complacency. Some of the people who leave the organization in the course of a year may include individuals who have been blocking progress.

How to institutionalize commitment and sustain momentum

There are various ways of protecting the change process and making it more robust. I call these *wiring the system,* because for me this conjures up the image of rigging up alarms that go off and attract attention when there is danger. Another way of thinking about this is that you are working to institutionalize the change process so that it becomes the new status quo. Here are some basic strategies.

- Maximize the number of involved stakeholders to reduce the dependence upon any single individual.
- Try to maintain widespread involvement and enthusiasm through a steady flow of information, recognition, and positive feedback—so that the process cannot be stopped suddenly. Like a riot, it may lose momentum and gradually fizzle out, but while it is in full cry, it is hard to shut down quickly.
- Use the power of routine. Many of us struggle to get up and get to work on time. But most of us have long since quit thinking about whether we have a choice, because it is part of our daily routine.

Exhibit 5.16 *Typical events that impact the process.*

Event	Likely occurrences during five years
The principal sponsor (the president) is replaced or moves on.	
A strong supporter within the top management team leaves.	
One of the key members of the support network, or among middle management, leaves.	
There is a recession or there are cutbacks affecting all organizations in your sector.	
There is a major union dispute.	
There is a fire, flood, or similar disruption that directly affects the organization.	
There is a war or natural catastrophe somewhere in the world that adversely affects your market or the cost of your operations (for example, cost of energy or materials).	
There is a regulatory or tax change that has major implications for your organization.	
There is a takeover or merger that involves your organization.	
Add your own items here. (Think about events observed during the past five years.)	
Total likely occurrences during five years	
Total likely occurrences during one year	

- Encourage commitments that are hard to back away from.
- Build feedback systems (like measurements, audits, and assessments) that are woven into the management system and will not quit. We all tend to react to feedback whether or not we like what it tells us—as long as it is there for us to see. The top management team is no different.

Here are some ideas for putting these strategies into practice.

• Get key customers involved, especially if they are on a similar journey. Key customers have power over your organization. They can insist that you do certain things, or take their business elsewhere. If the president of a key customer organization is a quality improvement fanatic, then your president may benefit in many ways from direct contact with this individual, and you may be able to arrange this.

• Get senior people invited to talk about the process outside the company—at meetings of their peers, to key customers, to the media. These occasions may seem little more than opportunities to pay lip service, but your colleagues may also become more knowledgeable through the preparation they have to do (with your help) and by talking with others. Quality improvement may even become a recurring talking point with these outsiders. This also associates your organization with quality, and primes people outside the organization to expect more information in the future about progress.

• Publish some information about the process in the annual report—preferably including quantified data like indices of employee morale and customer satisfaction, which a shareholder might relate to management competence and expect to see presented every year.

• Use the power of routine by starting up *processes* (which are ongoing), not *projects* (which happen once and are then completed). For example, have the team agree on the frequency of the newly designed customer satisfaction survey, and then entitle the first one the *The XYZ Organization* Annual *Customer Survey*.

• Use the power of routine in a similar fashion with everything you want to see happen more than once. Try to get timetables agreed upon well ahead for employee surveys, meetings of the Quality Council and the support network, recognition events, visits of senior people to key customers, and so on.

• Include in the plan a timetable of events that will provide positive feedback to contributors and sustain enthusiasm. The team cannot say "thank you" too often or show too much appreciation for the people who are making the organization successful. People will learn to expect recognition—and why not?

But more important, people will begin to view as a right the opportunity to make suggestions, to contribute ideas and see the best ones implemented, to get involved in improvement efforts that make things better for everyone. If these "rights" are ever withdrawn, management will hear all about it.

• Write an annual report, setting out accomplishments related to the change process. This is a feel good exercise, but one that also has real value. Change takes place like a gradual building process, and as each piece of the puzzle is put into place, it is quickly taken for granted and forgotten. A full inventory of achievements to date is usually a pleasant surprise. It can encourage those who have put in the effort to make things happen, especially when some are getting weary from efforts sustained over a long time.

• Get the planning and review process for quality improvement fully integrated into the normal management system so that it is not a separate limb that can be easily eliminated.

• Schedule regular audits or assessments. We have already discussed the use of an assessment during the start-up process. Periodic assessments are also a powerful mechanism for keeping the change process alive and on track. If the management team will commit to a broad-based assessment of the organization on a regular basis, this will surely help sustain the change process in the long term—perhaps more than any other single decision.

Key success factors

Have you covered the bases? Exhibit 5.17 provides a checklist of key success factors for the transformation. All are issues that need to be reflected in the plan.

The power of visualization

If you can dream it, you can do it.

—Walt Disney

Here is a very last test of the plan. Can you visualize the rollout, step by step? Can you visualize the way that your colleagues will step up to their leadership role, and the ways in which their actions will affect their people? Can you picture the way in which the process-improvement teams will move from faltering first steps, to achieving important changes to the way that some key processes work? Can you see how the activities you have planned will result in tangible gains that begin to justify the effort involved? Can you see who will report these successes and how?

This first plan, and these first steps, are a learning exercise for you and your colleagues. You cannot expect everything to happen

Exhibit 5.17 *Key success factors for the transformation.*

- An external (customer) focus
- Top management involvement and leadership
- Informed commitment to a long-term effort
- Buy-in of other key stakeholders (for example, unions, board, regulators, champions of previous initiatives)
- A plan developed and owned by those who will implement it
- Coverage of both technical and cultural issues
- A clear, practical vision of the future desired state, with measurable objectives linked to the key goals of the organization
- Systems for measurement, follow-up, and feedback
- An adequate infrastructure—for deployment of responsibilities and for guidance and support
- Use of proven methodologies, tools, and techniques
- Education—both targeted (for early results) and universal (for widespread involvement)
- Planned opportunities for individual pride, participation, and fun

exactly as you hoped, so you must monitor events closely in order to see what is working and what is not. If you're not clear on what you expect to happen, how can you know whether it's working or not? *The more clearly you can all visualize the rollout of the plan, the more confident you can be that the plan makes sense—and the quicker you will see when something is not working.*

Pizza time

At last, you are ready to begin the implementation of the plan! However, pause to recognize that in getting this far, a lot has already been accomplished, and you can take some personal satisfaction from this.

The assessment and planning process has involved your colleagues, increased their understanding, and even excited some of them. The top management team's commitment to change has been strengthened.

The change process has already been started within this small circle of key people. The next steps will simply widen this circle, to eventually include the whole organization.

You and your colleagues are now going to share a great learning experience over the next year or so. You are ready to start eating the elephant. OK—the whale, whatever.

6

Managing Change

To improve is to change; to be perfect is to change often.
—Winston Churchill

CHAPTER CONTENTS

Organizational change portrayed as a journey together of many individuals, each traveling the journey for their own reasons.

- The personal journey of the individual
- Reasons to resist change
- Reasons to go along
- Working through change—the personal struggle
- How change affects the individual
- Helping people through change
- The journey together of the group
- Typical patterns of behavior
- Typical patterns of progress over time
- Guiding principles for managing change
- Dealing with setbacks, slowdowns, and uncertainty

As implementation of the plan gets under way, many more people are going to be affected, and not everyone is going to welcome the changes. In fact, there are certain to be pockets of entrenched resistance, as well as groups of enthusiasts. Even those who welcome the changes are going to find the process stressful and draining at times. Everyone will have moments when they ask themselves "Why did we ever get into this?" This is all entirely normal; in fact, it is unavoidable.

This chapter explains what is going on—how change affects the individual, and how this in turn affects the organization's efforts to accomplish change on a large scale. This understanding of the dynamics of change will prepare you for the events ahead. Finally, we will suggest some guiding principles for managing change and some ideas for dealing with setbacks.

How the West was won

Here is a perfect metaphor for the process of transforming an organization. Picture the trek of a band of pioneers into the West. We see them on the trackless plains in their wagons, weary after months of travel. Conditions are hard, but they are sustained by a vision—the dream of a better life ahead. Some who lost faith in this vision have already been left behind or buried. Each individual carries mixed emotions in his or her heart—anticipation of better times, better food, a chance to rest awhile—but, also the regret of a life that has been left behind and is gone forever. They are also sustained by the certain knowledge that they cannot stop and they cannot go back. They must reach the mountains before the passes are closed by snow, and they cannot survive for long on this barren plain. They must keep moving.

Although they travel together, this is a different journey for everyone. For the young in spirit, unencumbered by fear or nostalgia, it is an adventure—full of hope and discovery. For some, who have already lost everything—possessions, family, friends—it is a forced march, only slightly preferable to death. For some it is a mission, a duty—to give their young the chance of a future denied them by fate. Some travel alone in their minds although surrounded by others. Others draw strength from helping others—cajoling, teasing, urging them on.

Each person travels on his or her own journey in his or her own mind, looking forward in hope, or looking back in regret. But the group as a whole cannot succeed unless each person will contribute, each in his or her own way. Some contribute brawn and endurance—others contribute courage and good humor. Some know how to mend wheels and tend to the horses—others know how to forage for berries and to trap rabbits. Some know how to tend to broken bones—others tend to broken spirits. This is the true spirit of the West—a journey made together in hope.

There seem to be some laws—rather like the laws of nature—that determine whether these pioneers will succeed on their journey.

- They need a clear shared understanding of where they are going together. Each person's dream of the life ahead is different. But the group's goal is clear—San Francisco.
- They need a code of conduct. If one individual hogs the water, others will go thirsty. If someone gets ill, or a wagon breaks down, the others must help out, because everyone will need help at some point, and this understanding binds the group together.
- They need leadership. This is a shared responsibility, but someone must carry the main burden. This individual must be the symbol of the group, the person who enforces the code of conduct by example and by moral force, inspires confidence and hope, and reminds them of their goal when spirits are flagging.
- They need to have ways of making the big decisions that affect them all. What can be done if the planned route proves impassable? A good leader will lead the group to shared decisions when there's time, and provide quick decisions in an emergency.
- They need a plan for how to get where they are going—which trails and passes they will use, during what months they will travel, who is contributing what equipment and expertise.
- They need guidance and know-how. A map is of no use without a guide. They need someone who can interpret the map and figure out alternative routes when necessary—preferably someone who has traveled this journey before. The guide also knows how to survive in territory unfamiliar to the others,

how to jury-rig a wagon wheel, how to find water when there seems to be none.
- They need equipment and tools. Wagons, horses, provisions, medicines, cooking gear, water containers—nothing essential must be forgotten, everything surplus must be left behind.

If any of these elements are missing, the chances of success are slim. Those pioneers who set out without the required preparation, or who could not learn to pull together, often marked the trail with their graves.

The West was won, not by haphazard risk-taking or self-centered individualism, but by courage combined with careful planning and organization, and a willingness to work together for a common purpose. The same qualities will be required for your organization to succeed.

CHANGE AND THE INDIVIDUAL

In this book, we are looking at a journey involving many people—the transformation of an organization. But there are many other smaller stories woven into this—the personal experience of each individual involved. We need first to understand the individual's perspective, because the group cannot go faster than its individual members.

Starting the journey: the personal decision

Change in an organization begins with the individual—with you and others like you. It is a personal journey, an experience that unfolds in your own mind—yet you cannot go alone. It's a journey you have to share with your colleagues. Because you work together, you have to travel this journey together.

The Pioneer trek was the big picture. Now we want to consider what the journey feels like to the individual—the leader, the scout, the head of a family.

The company heads for a better way.

Picture the scene when the president of an organization first begins to share her ideas for change with the top management team. "Why don't we aim to become a great organization by working more as a team, really focusing on the needs of our customers, and by giving our people the knowledge and tools they need to do the job?"

Here are some of the immediate reactions—although these may be thought and not spoken.

- "Sounds like *empowerment!* There's no way I'm going to let my people start calling the shots—after all the years it's taken me to get to this position. What are senior managers for anyway? Include me out."
- "Sounds great. We should have begun to think like this years ago. When can we start?"
- "Oh no, not another crazy idea! The last president had one every month. She's been smoking dope—or been to a conference."
- "I'd love to see this happen—but there's no chance. She'll never get this bunch to cooperate on this one. I don't want to be the dummy out in front like the last time."

- "Will this upset the superb new unit-pricing system we finally finished implementing last week?"
- "We should stick with what we know works. It's time to clear out some dead wood and show everyone we're serious about getting results."
- "Sounds like an opportunity to do a ton of training! We could have the grandest training program in the land. Lead on!"
- "Next we'll be having group love-ins and discussing our innermost feelings. I hate it when people bring up emotional stuff at work. Let's just get on with running the business!"
- "Our only problem is our margins—fix that and we're home free."
- "How is this going to fix the defective woggle-sprocket that's killing our sales right now?"
- "Can we break for lunch soon? I'm dying for a smoke."
- "Good for her! This is a courageous move and I think she might pull it off. Include me in."

These emotions begin to erupt as soon as some change becomes a possibility.

Your colleagues in the management team may have experienced reactions like this. If so, they have since worked through them by establishing a shared vision and goals, by agreeing on a code of conduct, and by working together on the plan. Most of them are now convinced that this is the way to go, that they can succeed, and that the journey will meet their personal needs.

So the top management team has already been through part of the personal journey—it has had the opportunity and the time to come to grips with the idea and to make it their own. Probably no one else in the organization has had this opportunity —you are about to spring the whole thing on them out of the blue. What type of reaction do you expect!

You need to recognize that

- No one is going to come on this journey unless they understand why. They may give the appearance of coming along, but there's a world of difference.

- People need time to understand why the journey is necessary, and to accept this. Why do they have to leave behind the familiar ways?
- People need to understand how the journey will be accomplished.
- People need to have input to the plan, so that their needs can be taken into account.
- People need to make their own personal plans for the journey.

Let's explore the reasons why an individual may or may not choose to come along.

REASONS TO RESIST CHANGE

The power of the status quo

Change has no constituency.

—Jack Welch, President and CEO,
General Electric

Most people in an organization have some investment in the status quo—there are some things they like and don't want to lose. These things may be as simple as the routine and camaraderie of meal breaks with the same familiar co-workers. For some individuals, simply moving to another work group may be a real wrench. For other people, change may threaten their hard-won empires or their cherished personal ambitions.

Most people also have some discontent regarding the current order—there are things they would like to alter. But those who benefit most from the current situation are those in power. They have the most influence, the most to lose, and the strongest motive to keep things pretty much the way they are.

Whatever the situation, you can be quite certain that many people in the organization, including a proportion of the management team, will oppose the very idea of any real change in the status quo. You would be naive to expect otherwise.

The key point is that *the reluctance of individuals to embrace change is a normal and unavoidable part of human nature.* A boat does not move through water without creating a wake, and serious change does not take place within an organization without creating mixed emotions among those involved.

Fear of the unknown

The most powerful hold of the status quo stems from the universal human desire for predictability and a sense of control over our lives. Leaving the familiar behind is inherently unsettling, and involves some risk. Because of this, we all find change stressful. When we realize that we need to give up some of our familiar routines or comfortable certainties, we tend to have second thoughts.

For this reason, even when people in an organization do fully understand and accept the need for a change, and know how the changes will occur, there will still be a clinging to the old ways. This opposition springs from uncertainty about the as-yet uncharted future—in short, from fear of the unknown.

Nostalgia for the past

Sometimes a reluctance to change arises from a feeling that something almost sacrilegious is being proposed—like desecration of the burial grounds—and this can arouse strong emotions. The holy ground that is at risk may be some basic principle laid down by the founder of the organization, a tradition that has been maintained for many years, or some symbolic link with the past.

For example, when Jack Welch was radically reshaping General Electric (GE) during the 1980s, the one business deal that generated the most emotional opposition was a fairly minor one in terms of its impact on jobs or the company's finances. This was the sale of the appliances division, whose toasters and irons carried the GE logo into millions of American homes. This was viewed as abandoning the company's original heartland—a signal that nothing was sacred.

Here are examples of changes that may arouse such heated emotions.

- Abandoning the original core operation of the organization, even if this were the manufacture of buggy whips
- Changing a practice that is viewed as the signature of the organization, or the key to its past success
- Changing of a uniform or a logo that has historic associations
- Changing a policy that is considered an article of faith even though it is no longer relevant

The common themes here are loss of a link with the past, loss of identity, changing of the value system. These are the landmarks and the symbols of home. Do we really have to leave them behind?

The concerns of different stakeholder groups

Let's consider the ways in which this journey may be perceived by the individuals within various stakeholder groups. It will soon be clear why some are reluctant to abandon the status quo and trek off toward some vision that they fear may prove to be just a mirage.

Top management

Top management usually gains or loses directly from the success or failure of the organization as a whole. Why on earth would these people resist change for the better?

A primary reason is internal competition, and, hence, politics—the inevitable result of objectives and reward systems that pit people against their peers. Most top managers did not get where they are by being fainthearted flowers. Like you, they are hardworking, smart, and ambitious. If the system calls for competition among peers, so be it—that's the game one has to play.

With experience, one can become pretty good at this game, and this is hardball—the big league. So, the struggles for power and turf, which go on to some degree at other levels, are often magnified at the top. In an environment of intense internal competition, the greater good of the organization is no one's top priority.

Another reason is short-term pressures. Top management people in most organizations are under intense pressure to perform in some dimensions. In a private sector company, these pressures are usually short term and strictly output oriented (see Exhibit 6.1).

When the top person's job security is dependent upon sustaining the stock price (or some similarly capricious measure) quarter by quarter, then of course some decisions are taken that compromise the long-term future of the organization in order to show a short-term gain on paper. Everyone knows what is happening, but until the systems that create this intense short-term focus are changed, everyone is compelled to play the game of short-term results—or suffer the consequences.

The president may have cause to wonder, "These improvement efforts clearly won't affect this quarter's results, and will have little effect on the year-end figures. Will I still have a job by the time we do start to see some payback?"

In other sectors, such as government, the pressures on top management may be even more counterproductive. For example, there may be demands to fulfill the campaign pledges of political

Exhibit 6.1 *Concerns of senior managers about short-term focused leadership.*

"His focus is exclusively on the stock price. In fact, he checks it several times a day. I guess this is not surprising as his bonus is based solely on this."

"He has no real long-term strategy—only to keep cutting costs. During the first year this was easy, and the results looked good. The second and third years were harder. By the fourth year, we were in a downward spiral—hacking away at important programs, cutting services that customers needed, mortgaging our future."

"He commissioned a high-tech control center, from which operations right across the country are monitored on an hourly basis. He holds a review meeting in there with his top management every morning. Now he knows what is happening every day. But where are we going? What is supposed to happen tomorrow, next month, or next year? No one knows."

masters, although these promises were made without the benefit of all the facts.

For many a president, it would be a dream come true to be able to forge the top management group into a cohesive team, pursuing long-term goals and working together for the benefit of the whole organization. By changing the system of management, and thus changing the pressures on people, it is possible to make great strides in this direction. But do not expect this type of transformation to happen overnight, and bear in mind that the external pressures will not go away.

Middle management

Middle managers generally have the least incentive to come on the journey, and the most reasons to stay put. However, until these individuals buy in, you cannot expect frontline employees to pay much attention to what top management is saying. Frontline employees cannot even participate safely—and have no reason to get involved—until their immediate managers and supervisors understand and actively support the change process.

Middle management has a lot to lose. Middle managers may be the best and the brightest, who have worked diligently for years to win their promotions. Now they may face being deprived of their just reward. Perhaps it seems to them that the more successful they are in pursuing this new approach, the quicker their jobs will disappear.

Sometimes significant numbers of middle management jobs are on the line. This often happens when a top-heavy organization struggles to shed layers—perhaps without much thought or concern for the human beings who make up these layers.

Even when middle managers' jobs and salaries are not at stake, their perceived power and status may be. Why should they give up the habit of handing out orders and having others respond, when they have worked so hard to secure this privilege? Some managers cannot face the thought of trying to manage in a more participative fashion, or don't believe that they could learn how.

If they should come to enjoy the new style of working, what about the loss of the trappings of power? What about the different types of role that an experienced person may be asked to play; for example, as an educator or advisor? What will their families and friends think if they no longer have an impressive title, a reserved parking space, and many people reporting directly to them? They may fear losing the respect of others who are close to them—undermining their own self-respect.

Perhaps every one of these managers knows that the survival of the organization is at stake, but they want solutions that don't turn their own lives upside down. And why should these people suffer for doing just what was asked of them? They didn't devise this unwieldy top-heavy structure, or invent this heavy-handed management style. They were taught this approach, by example and in the classroom, and they were rewarded for applying it. They were praised for putting out fires and being tough to get things done. They were given pay raises for enlarging their departments, and promoted for saying what the boss wanted to hear.

It does seem rather unfair to ask them to change now, doesn't it?

Frontline employees

Regular frontline employees are generally eager for change—once they understand why the journey is necessary, how it will be accomplished, and how they can participate. However, this is not possible until some significant changes have taken place at other levels in the organization—and until their own immediate managers are also fully on board.

Regrettably, to the frontline employee in too many organizations today, the working environment feels almost like a form of industrial apartheid. This situation is not intentional, and it may not be as marked as we will describe it here, but it is the common result of a combination of factors.

- Unconscious negative assumptions about frontline people's motivations and capabilities
- An ingrained pattern of learned controlling behavior among managers
- Poor communication, and little personal contact, between frontline employees and other levels
- Lack of attention to systemic problems that create barriers to frontline people performing their jobs

As a consequence, frontline people often have little reason to think well of the organization.

Frontline employees generally take the brunt of many of the problems caused by the organization's faulty systems. While their supervisors are in meetings studying charts of service failures, the employees are being chewed out by frustrated customers. While those in charge debate which department should pay for some costly debacle, employees have to tackle the dismal task of redoing or scrapping their own work, which is unsatisfactory for reasons beyond their control.

Too many organizations unintentionally disenfranchise and demean employees. Employees are treated as a kind of subclass—people who are not supposed to think, who are given little or no say in how the organization is run. Often no one even explains to employees the rationale for decisions that have a major effect on their lives.

It is often unconsciously assumed that employees are untrustworthy, or at least not capable of making wise choices about how to employ the organization's resources. So while millions of dollars are flowing through the system—and so much depends on the diligence of the employees' efforts—they often do not have the authority to spend a single cent on their own judgment and initiative. Who would treat their own small children with so little respect?

So, employees are often not very satisfied with the status quo. But they often reject new ideas that come down through the management hierarchy, and show anger, frustration, skepticism, and apathy. Why is this, when they have so much to gain?

"Here we go again." The first reason is that employees often feel that they have been fooled many times before into believing that the organization was sincere about some new policy or direction, only to find that it was a passing fad or a kind of bizarre ritual.

They may remember a previous directive that seemed to signal a change in priorities. Here is an example: "Any operator who observes that the outgoing product is defective in any way should immediately press one of the red buttons which stops the line." Does anyone remember the name of the nice young rookie who took that policy seriously? Did he find another job somewhere?

Employees would love to believe that the system will be changed, that the problems which are beyond their control will be tackled, that things will get better. But, based upon their experience, they will naturally exercise some skepticism and caution.

"No one asked my opinion." Employees reject new ideas because, based upon experience, they expect changes to be imposed on them by others, perhaps accompanied by some token consultation. This is deeply threatening. When others have shrunk your working world to a tiny place—a few square feet of floor or desk space, a few short breaks to rest, eat, and gossip—you will defend this little world for all it's worth, even if you don't like it much. It's all you've got.

Picture the situation in which you have been doing the same job for more years than you care to count. During the first week you observed that many of the rules, and the ways in which the work is organized, are dumb and wasteful.

However, your opinion is clearly not valued. In fact, your coworkers cautioned you from the start to keep quiet—not to do anything that appears to challenge the knowledge or the authority of those in charge. You certainly cannot get anything changed. To cap it all, every so often some fancy idea is forced on you, which is supposed to make you a better worker. Just who do they think they are?

"Just give me my paycheck and I'm out of here." A third reason why employees are reluctant to sign up is that people who are not allowed to become engaged in their work become resigned to this, and detach themselves, both mentally and emotionally. If they are required to do only what they're told, they quit thinking about the job. If they don't know how the work should be done in order to be perfect, they quit caring. If they are not respected, they lose some of their self-respect and pride of workmanship. And when employees are deprived of any opportunity to use their initiative, they learn to be passive in the workplace.

People subjected to this kind of regime for much of their working lives cannot be instantly switched on again. In their private lives they may develop outstanding abilities in sports, music, or art, but the workplace is a dead zone for them—a numbing experience to be endured so that the rent can be paid or the family's needs met.

When these employees are offered more responsibility, their first response is that these are management's tasks—they don't want them. When asked how things could be improved, their first

Exhibit 6.2 *Neglecting employees as a resource.*

> Comment overheard in the cafeteria: "Managers are going crazy trying to figure out why there are so many errors occurring in our section. Why won't they ask us?" (Said in a tone of sincere bewilderment and hurt.)

concern is often not the product or the customers, but the repulsive washrooms, the worn-out work stations, the lack of parking spaces for them. The unspoken message is "We don't trust you. Show us that you care about us, then maybe we will think about putting ourselves out to help you." Who can blame them? To reengage the talents and energy of these people requires a demonstration of management qualities that are often undervalued—such as honesty, respect, patience, and persistence.

I hear someone say "My organization is quite different—we treat our employees really well." Well, that's great to hear, but what has been done to find out whether your employees agree with this rosy picture? Remember, Marie Antoinette thought that the peasants had cake. If the above description of the current reality seems too harsh, you don't have to agree with it or accept it; but, if you don't have data, this opinion is as good as yours.

Other stakeholder groups

There are often other stakeholder groups to consider; such as the unions, the board (or an equivalent body), and shareholders. All of these are potentially supporters or opponents of the change process, and may need to be considered—especially if they are powerful or influential groups.

Union representatives are a bit like the opposition in a democratic system.

- They are usually elected and must seek to satisfy their constituents.
- The system within which they work was designed for opposition and confrontation.
- They have little reason to trust the people in power, and many reasons, based upon experience, to mistrust them.

The union people are likely to see a quality approach as a threat to job security at the individual level—even though it may necessary to keep the entire operation in business. Where there is a union, they can and must be made partners in the process, but the concerns that flow from their natural role need to be addressed.

The board members may be concerned about the organization getting a reputation for methods that seem unconventional or extravagant. There is a theory that some boards would rather risk the company going bankrupt than do something that might be perceived to be stupid.

Shareholders may think like traders rather than owners, and take a short-term view of the company. If so, they may be preoccupied by the stock price and quarter-by-quarter bottom-line results. So shareholder reactions to this type of approach may not be entirely sweetness and light: "You're going to spend a lot of time and money on systems thinking and interpersonal relationship training? Hold on now, we cannot afford to indulge ourselves in abstract management theories, or to treat employees with kid gloves."

It's easy to see why individuals at all levels in the organization, and various other stakeholders, may be slow to recognize what a great idea the proposed changes are. All of these people have some good reasons for opposing change—reasons that make good sense for them in their situations. They will come along willingly only if they can see even *better* reasons why they should step out of their comfort zone, give up their familiar habits and routines, and venture into the unknown.

REASONS TO GO ALONG

Although there are always reasons for the individual to back away from change, there are also reasons why change may seem attractive. We will consider these in the context of a work environment in which people are not completely isolated—they do have working relationships with their colleagues, and are influenced by their colleagues' opinions. Whether it is through peer pressure or team spirit, individuals are influenced by the group.

At a personal level, each individual may be swayed by the following types of factors.

Hope for the future

- The desire for control over your own destiny—for example, the prospect of being able to tackle chronic problems that bother you
- The need for a sense of accomplishment and self-worth
- A sense of obligation to others—for example, the desire not to let down the team or the customer
- Love of learning—for example, acquiring new skills
- The influence of other respected team members—a willingness to emulate people who are role models

Discontent with the current status

- Dislike of the status quo
- Fear of impending disaster
- A current lack of personal fulfillment

For the initial decision to hold up and carry people through the transition ahead, three key factors must be present.

1. *Pain*—A compelling reason for change, such as a threat to survival
2. *Vision*—A clear and practical vision of the desired future state
3. *Next steps*—An understanding of the next steps required to progress toward the vision

If these factors are not clear at the outset, they have to be developed and communicated. They may be thought of in terms of an equation.

$$\text{Change} = \text{pain} \times \text{vision} \times \text{next steps}$$

Since the three key factors are multiplied together, if any one is missing, little change will take place. For example, if there is significant discontent with the status quo and a clear vision of the desired future, but no plan, method, or tools to get there, little progress will be made.

The decision-making process

When several people are involved who are part of a team, there are some conditions that need to be satisfied in order for each person to reach a decision to embrace change.

- The plan has to accommodate everyone's individual needs well enough to ease their concerns.
- Everyone has to accept the idea of giving up some things they like about the present situation.
- Everyone has to make an effort to see what opportunities exist—for themselves and for others—in making the changes.
- No one's motives can stand in isolation. The team usually has to execute the changes together, so it matters to each person what the others feel about it.

All of this takes time, a lot of sharing of information and concerns, goodwill, and a willingness to compromise for others' sake.

If most of the team embraces change, but some members do not, these reluctant travelers continue to look back and become a drag on the others. Their hearts are still back in the "good old days," and inside they are saying "You made me come here and I hate it!"

WORKING THROUGH CHANGE—THE PERSONAL STRUGGLE

When a commitment to change has been made in principle, that is not the end of the story—it is only the start of the journey. Now the struggle of working through the changes begins.

Stress

Why is it that reaching for an opportunity would be a struggle, especially when it didn't involve trekking across the plains, but merely changing work habits and the like? As a result of research, we now know a lot more about how change affects the individual. Here are some of the facts.

• *All change is stressful.* Studies have shown a strong relationship between changes in a person's life, and the subsequent likelihood of them becoming ill.[4] Perhaps this is not surprising.

But, the strength of this relationship is surprising. On average, an American has a one-in-five chance of being hospitalized during any given period of two years. However, if you have several significant changes taking place in your life (like a house move, a change of job, or marital problems) the likelihood of becoming seriously ill within a two year period rises to almost nine-in-10. In other words, where two out of 10 average people will become seriously ill, nine out of every 10 highly stressed people will suffer major health problems.

Another surprise is that positive events can be as stressful as negative events. On a list of 43 life events, the death of a spouse ranks as number one. However, two positive events—marriage and retirement—are not far behind, in seventh and tenth positions, respectively. Even Christmas and other holidays appear on the list as causes of stress. An individual who is promoted during a reorganization may be as stressed as someone who lost his or her job—or they may be even more stressed.

• *The reactions of individuals to change tend to follow patterns.* For example, here is the typical pattern of response to a major traumatic loss, such as the death of someone close.[5]

- Denial—"This hasn't really happened."
- Anger—"Someone should be punished for this."
- Bargaining—"If I promise to . . . give away all my money . . . never be bad again . . . can this be undone?"
- Depression—"Life seems pointless now."
- Acceptance—"Life goes on."

• *People need time to work through these adjustments to change, and there is no shortcut. They need to proceed at their own pace.* Adjusting to change is work—emotional work—and it takes time and personal effort. In the case of a major loss, it seems that people need to work through all of the stages listed, one by one. In the best case, they do so within a reasonable time frame, and reach acceptance. The worst case is that they get stuck somewhere—and perhaps never get beyond anger or depression.

• *Stress is not all bad.* A moderate level of stress is normal and healthy. This is characteristic of having some excitement or taking on a challenge that we relish. Either too much or too little stress is undesirable—such as when we face challenges that are overwhelming, or lack any stimulation in our lives.

What has all of this got to do with making an organization more productive, improving products, or serving customers better?

- Improvement within an organization *always* involves change that affects the individual. Sometimes it involves frequent changes over a long period.
- Changes that may seem minor to top management are often major and traumatic to the people affected. The level of stress induced is not a factor of the size or the cost of the change, but the extent to which it disrupts people's expectations.
- People who are overly stressed by change cannot give their full energy or attention to their work. Very high levels of stress lead to a range of emotionally charged reactions and dysfunctional behaviors among people who are usually balanced and rational. For example, during a major downsizing, many of the reactions to a loss may be seen—denial, anger, bargaining, and so on. Even those who keep their jobs suffer in this situation—they lose colleagues, familiar work arrangements, and perhaps their sense of security.

Phases of personal change

To understand what each individual is going through, we need some kind of models of how personal change takes place. Here is a brief summary of ideas taken from the literature on this subject.[6]

For any individual, change may be thought of as having three phases.

1. Endings—letting go of the past
2. Transitions—becoming reoriented and exploring new possibilities
3. Beginnings—setting out in new directions

For the individual, change *starts* with endings—or letting go of the past. Only when this is done can people begin to focus their energies and attention on the task in hand. Endings often start with some kind of problem, which forces change on the individual, or triggers them to seek a new beginning.

Beginnings are the taking up of new directions and opportunities. However, these often do not flow directly from endings—there is usually some period of transition.

For example, you may move overnight from one job to another, without any evident break or interlude. But at an emotional level, there is a period of transition. You may have spent a long time coming to grips with the realization that the old job is no longer satisfying. This is an ending. You may then spend considerable time trying to figure out what new direction you want to take. This is the transition. The job change might simply mark the end of the transition—a period of search for a new purpose and direction.

Alternatively, if your job change was a forced one, the move may be followed by a period of learning to accept that the old job is gone (an ending), followed by a new search for purpose and meaning in your new situation (a transition). Transitions are not comfortable places to be, although they are often periods of personal growth.

This is the typical pattern of individual change. The person who is in transition is uncomfortable—disorientated, disenchanted and disengaged. He or she may also have somehow lost sight for the moment of who he or she is. The person in transition is in a kind of limbo, looking for a way forward.

Helping individuals to cope

It is in everyone's best interests to ensure that each person is helped along the journey. The group cannot move fast with many of its members looking back in regret. As more people can be inspired to fix their eyes on the goal, the rate of progress and the chances of success improve.

Here is how people can be helped to cope with the personal transitions they are going through.

Provide information

Naturally, people should be given information about why change is necessary, about the desired future state, and about what needs to happen to get there. How do you find out what information people need? You ask them; and when you feel that you've met their information needs, you ask them again if you succeeded.

In addition to information about the desired changes, it may also help to give people information about the effect of change on them (see Exhibit 6.3). The common feelings of loss, anxiety, disorientation, and so on will be easier to bear if people are prepared for these, realize that these reactions to change are normal, and that they are not alone in experiencing such feelings. It may even make sense to link employees with counseling services, to assist those who are suffering the most.

Let people plan their own journey

Much of the resistance to change springs from the universal human need to have a sense of control over one's surroundings and one's destiny. You've probably taken pleasure in rearranging your home at times—redecorating, moving furniture around, putting up different pictures. What a different feeling if you lived in a regime where some government employees might unexpectedly arrive one day to redecorate and rearrange your home in a style dictated by the official master plan!

Exhibit 6.3 *Employees' information needs during change.*

When a manufacturing plant was planning the change from a batch to a flow system of production, they realized that it was important to prepare their people for the changes. So, in addition to classes on the technical aspects of the new system, they offered a one-day class on dealing with change.

This class became the most requested training offering, and was provided by popular demand for several hundred people. It was attended, not only by most people within the plant, but by many people from other functions on the same site. In an environment of continuous change, this type of information was appreciated and sought after.

People do not resist change so much as they resist being changed. It follows that individuals will cope much better with change if they have a hand in creating it. This is one of the reasons why it is so important to engage people and have them participate in figuring out what needs to be done and in making the changes happen.

There are issues where employees may have little opportunity to contribute—for example, determining long-term strategy for the organization. People, however, may have some sense of ownership, and cope more easily with the changes, if they at least understand why the changes are necessary and what has to happen. Once this background is agreed upon, understood, and accepted, people at all levels can play roles in figuring out how to accomplish the goals.

Problem-solving and process-improvement teams are excellent ways of accomplishing this, so that employees are involved in the design of changes to their own surroundings and work processes. People need to be able to make their own personal plans for change.

Another good example of this need can be seen in the common dilemma of how to flatten the hierarchy—thus eliminating many middle management jobs. One of the most effective and least destructive ways of accomplishing this is to involve the managers concerned in planning these changes. When this is done, and the need for the change is properly communicated, the most likely result is that

- Individuals work constructively and rationally on a plan to accomplish the objective—even though the outcome may be loss of their own jobs.
- There is an acceptance that the change is necessary and that the process was fair, so the outcome is accepted even if it is unpleasant.
- Those who do leave go with their heads held high, confident of their self-worth and their employability—not as rejects or victims. The immediate financial impact may be just as severe, but the psychological impact is less and they are much better prepared to deal with their situation.
- The organization retains the trust and goodwill of those who remain—as well as the leavers, who help shape the reputation of the organization and may even become future customers.

Show empathy
When we become concerned about something, it helps us to share our feelings—even when nothing much can be done to change the situation. However, this only works when we feel that the other person cares. There is nothing more certain to bring out the worst in employees than to impose the stress and anxiety of major change, and to simultaneously demonstrate a lack of concern for the impact on their lives.

There is more to this than just making arrangements that are considerate, such as providing relocation benefits and housing assistance when people are asked to move to another city. Well-considered logistics are helpful, but these are not a substitute for personal concern and consideration for the emotional upheavals that people may be going through.

There may be anger and frustration at being forced to choose between disagreeable alternatives (for example, relocate or seek other employment), even if everyone fully understands the necessity.

Exhibit 6.4 *What people in change want and often get.*

What they want	What they often get
Empathy • Listening • Demonstrating concern **Control** • Ability to influence events • Exploration of ideas for action, options, alternatives • Involvement in planning **Information** • About what is going to happen and why • About how change may affect their feelings and outlook	**Autocratic behavior** • Giving directions • Making decisions without consultation • Excessive secrecy **Avoidance** • Management "in hiding" as emotions boil • Management too busy to listen • Management dealing with tangible issues only **"Rah rah"** • Fire up the troops

Adapted from Harry Woodward and Steve Buchholz, *Aftershock: Helping People through Corporate Change* (New York: John Wiley & Sons, 1987), pages 62–63. Copyright ©1987 John Wiley & Sons. Reprinted by permission of John Wiley & Sons.

If no one has the time or is prepared to listen to these frustrations, even if little can be done to alleviate the causes, then employees' feelings about the organization will be soured and irrational. Destructive behavior is more likely under these circumstances.

It does help people to cope with change if others show concern. Co-workers can help, but management may be perceived to have created the situation, and only management can take some of the actions needed to ease the pain. So it is important that managers demonstrate concern—by listening sympathetically, and by taking appropriate actions when possible.

Exhibit 6.4 provides a summary of the needs of people in change (and what they often get instead).

WORKING THROUGH CHANGE—THE JOURNEY TOGETHER

Hopefully we now understand better the personal journey that each individual must make. So, let's stand back and look at the patterns of behavior that emerge when there are many people involved in change together.

The 20-percent rule of thumb

Change agents have a rule of thumb for the types of reactions you can expect within any group of people when they are beginning to embark on the quality journey. A few (say 20 to 30 percent) will quickly become committed enthusiasts and standard-bearers for the new order. A few others (another 20 to 30 percent) will never accept the changes and will fight tooth and nail against them. This campaign may be conducted overtly or covertly, crudely or with style—but it will be fought.

The remainder of the group will usually go with the flow; that is, their support (or resistance) will be more passive. They may be very vocal in supporting the program, or any other politically correct initiative. They may even truly believe themselves to be very committed, but their actions will demonstrate that they have other priorities.

No two situations are alike, but this is a typical pattern. If you can see a few strong proponents emerging in key places, then the process is working well. You should not expect more than this at first. Fortunately, a few committed enthusiasts are all that's required to get the process started—provided the leader is one of these.

You should not lose faith in your colleagues because of this experience of hidden opposition. We've looked at the many good reasons why some cannot immediately see why this change is such a great idea for them. Put in their shoes, you might well behave in exactly the same way.

Setting priorities

The 20-percent rule of thumb provides a clue to one of the key strategies of managing change. If there is any secret that can be really liberating for the Change Agent, it is this: *You do not have to spend a lot of time and effort on those who strongly resist change. You only have to help and protect those who want to change, so that they are able to succeed.*

Put another way, your job is not to plant the entire forest, row by row—it is to plant clumps of seedlings in a few hospitable places and to nurture them. As they mature, these trees will spread their seeds, and the forest will eventually cover the fertile land. The rocks will, of course, remain barren regardless.

This is a logical, effective, and responsible way of using your very limited resources. This does not mean that you can afford to ignore the existence of committed and influential opponents of change. You may have to find ways to prevent these individuals from sabotaging the process. However, once you have figured out who falls into this category, you should not waste time trying to convert them.

Typical early reactions

Here are typical events and reactions that you can expect to observe during the early stages—let's say the first six months or so of your start-up. The following reactions are typical during (roughly) the first year.

- Enthusiasm and involvement
- Open opposition, skepticism, and cynicism
- Apathy
- Lip service and backsliding
- Anger and frustration
- Malicious compliance and sabotage

Responding to typical early reactions

Do these reactions indicate that are we making progress or not? Here is a saying that provides a partial answer: "Problems worthy of attack prove their worth by fighting back."

Clearly, the enthusiasm of a few is a good sign, but most of these reactions are good signs in the sense that they indicate that something is happening. The worst possible result of your efforts is no reaction at all, because this would indicate that everyone simply yawned and carried on as before. This nonresponse might occur if the issues you had chosen to tackle were of little importance, or if no one was convinced of management's serious intent.

If everyone seems very enthusiastic, this would also be a suspicious reaction. Perhaps they don't understand what was proposed (including the amount of effort involved), or they are not being completely candid.

The reactions listed earlier indicate that people do feel that something real is happening, and are choosing either to get on board or to challenge this new development. Energy is being generated, and this can be tapped and channeled in useful directions.

Most of the reactions described call for responses—from the Change Agent and from other senior managers—to help the process along. Let's consider what some of these responses should be.

Enthusiasm and involvement

A few people will emerge who are prepared to accept that this initiative may be legitimate. These people are excited by this possibility, and want to help make the changes happen. They want to get going quickly, and may bombard you with requests for information and guidance. A few people seem to know instinctively what is required, and naturally take to the new approach. Some already have done, or attempted, something similar within the organization.

These enthusiasts are the flagbearers for change who make success possible. They need to be nurtured and supported to ensure that they succeed—and protected from opposition or sabotage by others. They also need to be recognized for their contribution, and held up as a positive example to encourage others.

There is another reason to stay close to these pioneers: Their enthusiasm and lack of experience may cause them to overreach. They may need special support to succeed, and recognition for worthy efforts if their initial accomplishments are modest.

If a team is not succeeding, there is usually some way of regrouping, restarting, or reframing the objective, to enable something useful to be accomplished. The aim is to ensure, as far as possible, that no group feels that it has failed—although the members might feel that they learned a lot from their initial struggles.

Open opposition, skepticism, and cynicism

Many people will dismiss the new initiative as yet another fad, following the well-worn pattern of many previous initiatives. Their contempt for senior management's inconsistency may be barely concealed.

Some open opposition comes from skeptics. These people can become the strongest possible flagbearers, but they are presented in disguise. They may exert considerable influence over their peers by virtue of their outspokenness. They may also be people who demonstrate the courage of their convictions.

Exhibit 6.5 gives an example of how an outspoken skeptic may be converted to an ally. This scene is a frequently repeated one, and it illustrates how management can demonstrate its commitment by a single highly visible decision, or demonstrate by a poor decision that nothing has really changed.

It is worth remembering that showing greater concern for employees and seeking their involvement does not mean throwing away all the rules. A construction company might involve employees in efforts to improve safety and may encourage them to use judgment and initiative in dealing with potential safety problems. However, anyone who refuses to wear a safety helmet still gets disciplined.

When opposition is carried to such lengths that it breaks necessary rules or endangers other employees' well-being, then disciplinary

Exhibit 6.5 *Skeptic turns enthusiast.*

> When GE began systematically to open up communications with its frontline people in the late 1980s, the following scene took place at one location.
>
> The plant was receiving screws that did not work well with the power tools. As a result the screw heads would break off, scratching the product and sometimes cutting operators' hands. For years, management had consistently refused or failed to get this problem fixed. During the "Work Out" session, a shop steward stood up and told this story, and explained in considerable technical detail what the causes were. As a middle manager there said: "This guy was a maverick, a rock-thrower, a nay-sayer. He wanted to test us, to see whether we really wanted to change."
>
> In this case, management passed the test. They asked the steward to lead the efforts to fix this problem, and flew him and his team to visit the offending supplier. The result? "The steward became a leader rather than a maverick. . . . Now we don't even have supervision in his part of the plant. He carries a two-way radio, and if he needs help he asks for it."

From Noel M. Tichy and Stratford Sherman, *Control Your Destiny or Someone Else Will* (New York: Doubleday, 1993), 202. Reprinted by permission of Doubleday, a division of Doubleday, Dell Publishing Group.

action is called for. This will often be welcomed by other employees. If there are unions involved, this is the type of situation where their support is essential.

A skeptic's opposition is founded upon sincerely held concerns. There are some others who will offer open opposition out of destructive malice or narrow self-interest. Such people are like vandals who need to be contained for everyone's benefit.

There are also other individuals, better described as cynics, who are simply chronic malcontents. You need not worry too much about these individuals. You can distinguish them from skeptics by the fact that their co-workers find them as irritating as you do.

Apathy

People may demonstrate complete apathy. They may ignore what is going on and show no interest or desire to get involved in making things better. This, too, can anger and frustrate their managers, particularly when they are making a genuine effort to empower people.

There is no instant solution to apathy. When managers encounter this reaction, they need to be patient and to nurture any signs of 'volunteerism' until the idea of getting involved begins to catch on. Managers need to keep talking with their people and listening for the issues that arouse emotion and energy. When they find out what these issues are, they must be prepared to take action that will demonstrate that they take people's concerns seriously. And they need to hold out to people the prospect of gaining some control over their lives in the workplace.

In a sense, the entire quality improvement process within an organization is a voluntary one. People can be encouraged to go through the motions, and they may accomplish some significant improvements in this way, and enjoy the experience. But we all have an idea at the back of our minds, like a switch indicating an unconscious decision, that we want to do something—or we do not.

For each individual, the improvement process really comes alive when this unconscious switch flips, and someone decides that he or she does want to be involved in making things better in the workplace. This is the difference between a volunteer and a conscript.

Lip service and backsliding

Some people will seek to give the appearance of enthusiastic support, but will fail to deliver. They will repeatedly miss commitments because of unexpected events and will plead for more time. They may be apologetic about it, but the unspoken message is that they have other, higher priorities.

Those who give lip service but don't deliver usually belong to the middle 60 percent—those who will tend to go with the flow. These people need to be held to their commitments and convinced that their actions will have consequences. These consequences may be pleasant, if they can emulate from the example of the enthusiasts, or unpleasant if they cannot.

In the early days, while everyone is still internalizing the ideas and experimenting, a more patient and laissez-faire approach is prudent. Perhaps the people concerned simply do not understand what is required, because the information they have received doesn't make sense to them. Or, perhaps they lack the skills to do what is required. Management needs to do a lot of listening.

However, as issues of understanding and skill are cleared up, and more enthusiasts emerge, there is a need to progressively apply more persuasive pressure to those who are not making a genuine effort. This process relies upon having clear tangible objectives and regular follow-up.

Holding backsliders to their commitments cannot be done effectively by the Change Agent—it is the responsibility of the line manager to orchestrate this. Only when people are convinced that they need to go in the new direction can the Change Agent help—by showing them how.

Peer pressure within the team is often the most effective mechanism at the line manager's disposal for bringing slow starters into the process. Enthusiasts, who are taking a risk or putting in extra effort to improve, can become quite angry at others who do nothing. As more people get involved, those who are opting out will feel increasingly isolated, and will be prompted to reexamine their attitude toward the changes.

At the same time, management has to be listening and looking for the barriers that are impeding these individuals. All the pressure in the world cannot make people want to do something when they don't understand why—or make them able to do something when they don't know how.

Anger and frustration

Some managers may become frustrated by the reactions of their people. Their people may be angry at management for yet again attempting what they assume to be some kind of deception.

Often, management's initial attempts to open up communications result in a deluge of vocal, emotional, and perhaps unreasonable complaints. Management may be frightened by a sense of the situation getting out of control—especially if this opening up takes place in a meeting that turns into a free-for-all.

Initial attempts to create two-way communications are often like opening a floodgate. If someone asks your opinion for the first time in 15 years, you may have a lot to say before you get it all off your chest.

However, managers are often unprepared for such a strong reaction to their early attempts to get feedback, and do not know how to deal with it. Their first experience may put them off the whole idea of more open communications with employees.

It is important to prepare managers for their new role, including situations like these. Efforts to open up two-way communication with employees often trigger such strong reactions. Managers need new skills to encourage employee participation, and to channel in constructive directions the energy that is released by doing this.

Skills training can help; for example, in listening, accepting feedback, encouraging diversity, and managing conflict. It also helps greatly if middle managers first see these behaviors modeled by top management.

Malicious compliance or sabotage

A few who are opposed to the change may follow orders in such a mechanical and unthinking fashion that the initiative is certain to go nowhere within their areas. Or, some middle managers may make it clear to their people that, regardless of what other messages they may be getting, business as usual will continue in this department—their people are forbidden to waste time on this latest craze, but told to "get on with their work."

It should be clear by now that there are many reasons why people at any level may sabotage the process, and it is dangerous to jump to conclusions about their motives in any particular case. You need to talk, listen, and probe enough to have a reasonable idea of what is going on, and then decide what to do—if anything.

In some cases, you may do very little, because this would be wasting time and energy that should be spent supporting the enthusiasts, and nurturing the undecided. We will review this type of decision later when we discuss setting priorities. In other cases, some kind of damage control will be essential, because influential and determined opponents can wreak havoc among those who are leading the way.

The first two years: Who does what

The following is a capsule of what typically happens during the first 18–24 months following the completion of the plan, focusing on the critical roles in sustaining the process. This illustrates how the leader of the organization and the Change Agent must both play their parts for the change to succeed. Other members of the senior management team can also do a lot to help, but some may still be opposing the process. Here is how the scene may typically unfold.

Supporting the pioneers

When the changes are initially proposed, you will soon begin to identify potential champions in the organization—people who are enthusiastic, who demonstrate leadership, and who will put time and energy into creating change within their own spheres of influence. These people are the forest seedlings—they must be nurtured and protected.

These early enthusiasts make everything else possible, and you must do everything possible to help these pioneers to succeed. Your main role is to engage the top management team in recognizing and encouraging these pioneers—as well as personally offering support.

In addition to direct recognition, you should ensure that these efforts are widely publicized. This ensures that others know what was accomplished and how. These early champions will persuade many of their peers to follow their lead, often more effectively than top management could.

It is often hard to know at the outset who will support the process. The most vocal opponents and skeptics may turn out to be the strongest leaders of change, once they realize that management is sincere and committed, while those who seem the most enthusiastic at first may simply be the most prone to pay lip service.

Encouraging the uncommitted

As the process unfolds, the uncommitted voters begin to realize that something significant is afoot. These people observe what the pioneers are doing, and begin to get more involved. The Change Agent (and the support network) may be overwhelmed by their requests for help.

As these latecomers achieve their first small successes, they take pride in their accomplishments, and they too are recognized for their efforts. Many now become active and supportive—to a degree. Most are essentially still waiting to see which way the wind will blow at the end of the day.

Weathering the storm

Over time—let's say 18 months—the realization begins to dawn that the old ways are being seriously challenged. The new approach, although not yet universally applied, has become widely accepted and is probably here to stay. This realization prompts the last stand of the old guard—those who are defenders of the old ways.

This is a turning point—and a time of crisis. Let's look more closely at what is going on.

- Some of the novelty and excitement of starting teams and learning new methods has worn off. Many teams have accomplishments they are very proud of, but a lot of hard work was required, and there is so much still to do. Some people feel the need to rest and recharge their batteries—they are heading for burnout.
- Memories of how bad things used to be have dimmed. There may be some unrealistic nostalgia for the good old days.
- Employees like having more say in decision making, such as how local budgets should be allocated, but are not so pleased about having to face up to the difficult choices that this entails. In the old days, being powerless also meant being free to complain, without any responsibility to think or to act.
- Some people are concerned about their jobs. Middle managers are concerned because employees are taking over much of what they used to do. Some groups of employees are concerned because the non-value-added work they do—such as rework, inspection, and progress chasing—is becoming unnecessary.
- Those who are fundamentally opposed to the changes have been mustering their forces, gathering evidence to discredit the new order, and preparing for a counterattack. They realize that the time to act is now or never.

Regardless of what the Change Agent does now, everything depends upon the leader of the organization. Will he or she be unnerved by the chorus of discontent, or will the leader reaffirm the chosen direction, remind people that the old status quo is not an option, and inspire them to go on? If the leader begins to show doubt or a weakening of resolve at this point, the show's over.

The Change Agent does have a vital contribution to make in order to ensure success at this moment—to prepare the leader well in advance for this type of crisis. The leader must understand that moments like this will inevitably come, though it may be impossible to predict when, and that these are the times when everything depends upon the leader's unswerving commitment.

If this battle is won, then most of those who have been uncommitted will now quickly join the winning side. They still need help to master the new ways, but now they realize that they can no longer just go through the motions and wait for this fad to blow over. However, this still leaves a few individuals who just cannot or will not get on board.

Finishing the job

In this scenario, there is one more task that the leader must begin to tackle now in order to complete the launch of the process. This is to begin to deal appropriately with any members of the top management team who clearly will not participate.

If some senior people continue to resist the new approach more or less openly, their presence will serve as a constant reminder to everyone that the management team is only partly serious about these changes. This will send out the following kinds of mixed messages.

- Management wants everyone to work as a team. But, if you don't find it in your nature to do so, that's OK too, especially if you are a vice president or a star salesperson.
- It is important to identify and manage key business processes. But, if you play golf with the president, you can always opt out by offering the view that you don't have any processes.
- All managers must treat their people with respect and give them the opportunity to help improve the business. But, if some long-serving managers find this too difficult to do, well that's "just life." After all, no one's perfect.

As long as this situation continues, there will be a constant risk of the enthusiasts losing heart and the whole organization backsliding. When the next change in leadership takes place, there may be an opportunity for a counterrevolution.

The leader must eventually confront this issue. The Change Agent should stand ready to help in this process if called upon, but it is the leader's job to tackle the problem. At a minimum, the leader needs to make it clear to these individuals that they are now out of line, that their behavior is not in the best interests of the organization, and that continued opposition will affect their career prospects within the organization.

The top management team also plays an important role here. If this group is becoming a cohesive team, then the refusal or inability of one or two individuals to get on board may put them increasingly at odds with the rest. Again, peer pressure may be the deciding factor for these individuals.

The ideal outcome is that those senior people who will not participate find a way out—perhaps opportunities within other organizations where their experience is valued and their style is acceptable. Their departure also clears the way for some new faces—people who are more in tune with what is being attempted, and who bring fresh impetus to the process. In this way, everyone wins.

The Change Agent's role in finishing the job

A caution is needed here about the Change Agent's role during this phase. As a change agent you are a helper, not a spy. You must avoid being sucked into the position of turning in colleagues. However, you do need to ensure that the leader is able to figure out for him or herself what is going on, without the need for spies.

When issues arise regarding individual contributions, the work already done to help top management function as a team will make a big difference. As the team members truly begin to act as a unit, it becomes more likely that such issues will be discussed openly and dealt with on the basis of fact, rather than by rumor and innuendo. Also, when the team is cohesive, peer pressure within the team becomes a more compelling reason for individuals to get on board—or get out. This situation illustrates another reason why it is important to establish tangible objectives for the senior management team, and reliable mechanisms for measurement, reporting, and feedback.

Consider, for example, employee morale and employee perceptions of senior management. These should be matters of record, based upon reliable survey data—not matters of opinion based upon anecdotes. There should be systems for obtaining input from various sources—for example, from peers, customers, and direct reports—in order to appraise the performance of senior people.

Holding onto the values

Finally, in dealing with these individuals—who are, in a sense, victims of the change process—the leader and the team should not forget about the values agreed upon at the outset. These recalcitrants cannot be allowed the privilege of opting out, but neither should they be denied the respect and consideration due to any employee. Right up to the end, there should be an openness to understanding better what is getting in their way and helping them to get on board.

GUIDING PRINCIPLES FOR MANAGING CHANGE

There is no simple formula that can be applied to all the situations that arise in trying to guide and sustain this journey. This chapter offers some ideas, some ways of thinking about what is happening, and some guiding principles. The rest is up to you. However, you can take comfort from the following facts.

- If you did your homework, most of the issues that need to be dealt with are already addressed in some way in your plan.
- You are not the first—many other organizations have already done what you are attempting.
- You can draw upon the practical experience of outsiders and other organizations to help you. We describe how to do this in chapter 8, "External Resources."
- You already have access to the most important sources of information to help you figure out what to do—your people and your customers.

Exhibit 6.6 *Guiding principles for managing change.*

- In working with individuals to win their commitment to change, start with their needs and ambitions—listen, don't preach.
- Work on building the commitment of the president and other key sponsors, and never stop reinforcing this commitment.
- Make the change process a team effort and, thus, ensure that everyone involved has the opportunity to take ownership of the process.
- Build partnerships that include all the key stakeholders: those who have the authority, the resources, and the expertise.
- In supporting the transition, set priorities and focus your efforts where they can be effective. Work with the enthusiasts who will lead the way and don't waste time trying to convert those who are lost causes.
- Strive for some (small) tangible early successes, and make the most of these through recognition and publicity.
- Ensure that those who are affected by the changes are involved in planning their own journey.
- Strive to act as a role model for others.
- Provide information to those affected about the need for change, the means, and also the effect of change on people.
- Listen, and offer empathy for the stresses people undergo during change.
- Celebrate progress, and make it fun.

In an earlier chapter we set out the key success factors for the change process. Exhibit 6.6 sets out some key principles for you—and for the top management team—to bear in mind as you go about your work during the transition. These are key success factors for managing change.

DEALING WITH SETBACKS, SLOWDOWNS, AND UNCERTAINTY

Learning in phases

We have set out the change process in a sequence, because this provides a logical flow—and this does portray accurately how one step tends to lead to another. However, this book would do you a big disservice if it left you with the idea that the process will unfold in your organization step by step as we have described it here—with an orderly progression from one phase to the next, and a clear grasp of what you are doing now and what you will do next. It is much more likely that

- Your learning will take place in phases, like moving from plateau to plateau.
- You will not be able to see very far ahead in terms of where you will have to focus your efforts.

At any given time, you and your colleagues will be preoccupied with mastering some particular aspect, such as empowering people, because of a conviction that "this is the biggest barrier to progress right now." You will not know that you are succeeding until you begin to see the desired changes in attitude and behavior take place. Then there will be a sense of excitement and accomplishment, and your colleagues will say things like "I never realized just how much we were trying to control every aspect of what our people do," or "I would never have believed that our people could get so turned on just by us getting out of their way."

These are reflections on what they have learned, and the learning has not taken place until the individual has been through the experience. So, no matter how many books and articles you read, no matter how many other organizations you visit and study, you can really only learn the process a bit at a time. It's like learning to ride a bicycle. You reach a stage where somehow, although you wobble, the machine miraculously stays more or less upright. You have learned to balance, and you say "Wow, this is what it feels like! I can do it now. It feels great!" Then you can think about the next level of cycling expertise—perhaps turns, or perhaps getting on and off without help—it's your choice. Wheelies? Perhaps a bit later.

So this is a voyage of discovery for everyone. You may know intellectually where you want to get to, and have some sense of what may lie ahead, but you end up feeling your way step by step. Your journey won't be exactly the same as anyone else's, and although others can help you will ultimately have to decide for yourselves where to focus your efforts at any given time.

Recovery from setbacks or slowdowns

Failure is the opportunity to begin again more intelligently.

—Henry Ford

Often, some early setbacks stop progress in its tracks, and it seems that the organization will revert back to its old ways. For example,

- Top management may become preoccupied with other issues, or lose heart as it realizes how much effort is involved.
- Important segments of the organization may reject the chosen approach, finding it unhelpful, too difficult, too different from their normal way of working, or too easy just to carry on as before.
- A new leader may come on the scene, eager to make a break with the past, with very different ideas about how things should be done.

What should you do? You may need to start over—at least in your own mind. You should go back to basics, and think through what is required to make the process work. *There is no failure mode that cannot be diagnosed by looking at the situation as if you were starting from scratch.* For example,

- When the chosen methods don't seem to be working, this is often the result of something important missing from the plan; for example, an imbalance of effort on hard versus soft

issues, lack of top management involvement and support, lack of tangible goals, or lack of recognition of initial efforts.
- The complacency that often follows some success will show up as a question mark over one of the must haves—a compelling reason for change. People may lose sight of the original reason for starting the process, or perhaps the success accomplished means that the original reason for change has gone away. Perhaps the near-term survival of the organization now seems more assured.
- Perhaps the original vision of the future has been largely accomplished, or no longer seems appropriate.

So, go back to basics, and review the checklists in this book that relate to

- Assessing the organization's readiness for change
- Key success factors for the process (elements to be covered in the plan)
- Guiding principles for managing change.

Also, review the final chapter on personal survival and growth. Perhaps the problem is not that the process is faltering, but that you are burning out, losing perspective, or expecting too much. Pay special attention to the following points.

• You need to recognize early that something is wrong. The earlier this realization dawns, the better the chances of an early diagnosis, some timely action, and a successful recovery without too much drama.

• You may need to reexamine the situation. Perhaps your initial assessment of the situation was inaccurate. For example, perhaps you misjudged the character of the president, or the position of some of the stakeholders—corporate headquarters, the board, or the union.

• You may need to dig deep to figure out what is happening in the process, and you may need help to do this. Every outcome has causes, and you need to identify them. You need to ask *why* several times, and dig down past superficial causes to root causes. You may not be able to do this well or quickly without some expert outside help—to provide a fresh viewpoint and a more objective perspective, and to draw upon implementation experience that you may not have.

• You need to build on what has been done. If the experience to date seems to have been disastrous, there should still be *something* to salvage, even if it is only the collective learning experience. Perhaps you realize now that the initial plan set out to fix the employees rather than the management system. "Boy, we won't make that mistake again!" you all say.

Hopefully, you won't give up because of that one (big) mistake. You may even find that this experience has strengthened, rather than weakened, the conviction and resolve of those who are committed to change.

• As a Change Agent, you need to learn from the experience. You shouldn't take on your shoulders the responsibility for everything that's gone well or badly. But there's surely something you can learn by reexamining your own contribution and by facing reality unflinchingly. If you feel that you messed up, be kind and forgive yourself—but try to identify precisely *how* you messed up, and figure out how to do a better job in the future.

Looking ahead: Expectations for the early years

This chapter completes a scenario covering approximately the first two to three years, as top management prepares to launch the process, create a plan, and deal with the ensuing turbulence of change.

If after (around) two years you have accomplished the following, then you have accomplished a great deal!

- Top management is still committed to this approach and is still working on it.
- There is an updated plan for improvement imbedded in the overall plan for the organization.
- Some significant behavioral changes have taken place.
- Some tangible gains have been won.

Note in your calendar some milestone—like the date of the first momentous meeting when the top management team decided that something must be done. Each time this anniversary comes around, if the change process is still alive and you are still in one piece, then you are succeeding and it's time for pizza!

7

Internal Resources

I get by with a little help from my friends.

—The Beatles

CHAPTER CONTENTS

- The pros and cons of the Change Agent being responsible for a staff or line function
- How to establish, sustain, and develop a network to support the change process
- How staff functions are changing
- Executing this transition
- The relationship between the support network and the staff functions

THE CHANGE AGENT'S BASE OF OPERATIONS

The issue often arises—should the official Change Agent be one of the existing staff function heads, such as marketing, quality, or human resources? One of these may seem a natural choice. Sometimes someone who is currently in a line management position may want to take on the role of Change Agent as an additional responsibility.

As the official Change Agent, you may, through choice or circumstance, operate from any of three different bases: from a staff function, from a line function, or as a free agent—someone who is not responsible for a department. There are advantages and disadvantages to each of these arrangements, which may be summarized in Exhibit 7.1.

Let's elaborate on some of the key issues involved in arriving at this decision.

- Can the current function head acquire the knowledge and skill sets required? Is the quality person open to learning a lot more about the human dimensions of this process? Is the human resources person open to learning a lot more about systems and processes? Does the operations person have the time to become a student again?
- What type of relationship with others does the ongoing functional role call for? Is this compatible with the relationships required to fulfill the role of change agent? For example, if the quality function is required by external regulatory bodies to execute an independent watchdog role focused on technical issues, will this get in the way of developing a supportive relationship with colleagues and dealing with management issues?
- What can the individual stop doing to release the time and energy required for these additional responsibilities?
- How will senior management signal to others that some significant change has taken place?

Often there is an assumption that, because quality management principles lie at the heart of this process, the head of the quality

Exhibit 7.1 *The Change Agent's base of operations.*

Pros and cons of combined responsibilities			
Issue	**Impact of other responsibilities**		
	Line	**Staff**	**None**
Time available for becoming an expert, helping colleagues, sustaining the support network of other change agents.	Con	Con	Pro
Ability to devote all your energies to the Change Agent's job.	Con	Con	Pro
Ease of building trust among colleagues—consider baggage, allegiances, rivalries.	Con	Con	Pro
Likelihood of problems in your own function absorbing all your energy and, thus, impacting the entire change process.	Con	Con	Pro
Ability to communicate that something has changed.	Con	Con	Pro
Clarity regarding role—consider risk of confusion between different roles.	Con	Con	Pro
Feeling of security—if the change process dies, do you still have a job?	Pro	Pro	Con
Ability to work within your own function—as a relief from working without authority to influence others.	Pro	Pro	Con
Ability to experiment and try out new ideas within your own function first.	Pro	Pro	Con
Ability to use your own function as a demonstration of how to implement the process.	Pro	Pro	Con
Credibility with line management peers as someone who deals with similar problems.	Pro	Con	Con

function is the person who should automatically take on this role. This individual definitely has a critical role to play in supporting the process, and may indeed be the ideal choice, but this should not simply be assumed. There is no one right way, and the attributes of the individual concerned are critical. If the head of the quality function is to become the official Change Agent, the following issues also need to be addressed.

- How will the senior management team avoid perpetuating the traditional belief that the quality department alone is responsible for quality; that is, that nothing has changed.
- How will the various resources required to make up the support network be marshaled? It is unlikely that the quality function (or human resources, or any other staff function) on its own can provide an adequate support network for the change process.

Whatever approach is adopted, it is essential that the individual be able to give the time required to do the job properly. Failing to invest the time and effort needed is a recipe for failure. Coming up with a half-baked approach to these issues may doom the process before it is even begun.

THE INTERNAL SUPPORT NETWORK

In chapter 4, "Preparing to Launch the Process," we described briefly the infrastructure required to support the change process. This includes

- The hierarchy of line managers who drive the process by disseminating responsibility and accountability
- The network of support resources who support the process by disseminating education and guidance

Both aspects of this infrastructure are essential in order to carry out the transformation. If either is missing or inadequate, the process will fail.

The tasks of the support network may eventually be subsumed into normal operations, but this network is critical in the early stages. People cannot embrace new ways of working if they do not know how.

In this section, we will focus on the support network. As the official Change Agent within the top management team, it is your responsibility to establish, sustain and develop this network. You are the network leader.

Purpose and structure

The support network has the following purposes.

- To provide education, information, and guidance on how to implement a quality approach
- To offer moral support and encouragement to those who are making an effort
- To ensure recognition of successes and worthy efforts
- To provide role models for behavior and for methods
- To serve an integrating role, helping to build communication and cooperation between departments

These purposes are achieved by establishing a loosely bound web of change agents throughout the organization. These are counterparts of the official Change Agent, who works within the top management team. These people work mainly within their own natural work groups, but may also provide help to other groups. This network is not a department or a function. It is an alliance of like-minded people who continue to work within their own areas, but who cooperate to achieve a common purpose.

These individuals are typically responsible for tasks such as development and/or delivery of training, facilitation and coaching of quality improvement teams, or coordination of projects like setting up a customer survey system. They often have other major responsibilities, and their support role is a part-time task.

In a large organization, more resources will be required and the structure of the support network may be more complex. For example,

- If the organization has independent or fairly autonomous divisions, each may have its own Quality Council and its own change agent(s).

- Some full-time corporate resources may be required, in addition to the Change Agent, to undertake tasks like education for people within the network, selection and development of educational materials, and research. This corporate team is managed by the Change Agent, and rarely comprises more than a handful of people.

Why a network is effective

This type of network is highly effective because, unlike conventional staff functions, it is not seen by line management as a separate department. It is seen as Bill or Mavis, who are members of the group—one of us.

These people always seem able to help get things done—with an idea or a smart technique, with encouragement, or just by digging in to help. They're also great at finding someone with the expertise to help with a sticky problem and knowing what's going on in the organization. Why, it's almost as if they had access to some fantastic . . . network.

This network has the power of a team of volunteers. The members are drawn to a common purpose, they cooperate rather than compete, and they are prepared to put their hearts and souls into the job because they believe in it. This network also provides excellent development opportunities, and a training ground for people who will in time take on leadership roles at all levels throughout the organization.

Establishing the network

The network may be loosely bound, but establishing it is not a casual affair. This task calls for a well-thought-out strategy. Here is what your approach might look like.

- You ensure that the top management team understands the necessity of assigning appropriate support resources and the benefits of a network approach. (The alternative is to give you the budget for a large staff group—an approach that you do not want.)

- You communicate your wish to have a say in the choice of network members, and offer your colleagues help to find suitable resources within their departments.
- You agree with the team on the desired characteristics and the core mandate of individuals within the network, as well as any specific assignments (for example, a full-time process management trainer/facilitator).

  You need people in the network who
  - are constructive
  - have the best interests of the organization at heart
  - have the trust and confidence of their boss and their colleagues (this is critical)
- You ensure that information about job opportunities within the network is widely communicated. Publicizing the mandate allows suitable people to volunteer, thus providing candidates who are interested rather than conscripted. Some of the people who volunteer may be *very* capable—people that your colleagues might not choose for this role because they are too valuable. If the organization does not already have some system for communicating job opportunities, this occasion could provide the vehicle for piloting a formal job-posting system.
- You scout for talent. You have already asked your colleagues to let you sit in on some meetings with their people, and this allows you to spot likely candidates. Consider those who championed previous improvement initiatives, and don't forget the people on the team that performed the assessment.
- You propose candidates to each of your colleagues, and discuss their preferences. You should avoid assembling a team of clones—especially clones of yourself. Value diversity, and welcome people who bring very different strengths and outlooks. This makes for a stronger team—one where people learn more from each other. Don't be afraid to include people who are clearly a lot smarter and more experienced than you are. You should be delighted if you can get such talent on your team.

As you begin to get these people signed up, you can begin to pull gently on the threads of the network.

The Change Agent's role as network leader

Your ongoing task now is to

- Establish and maintain alignment between the vision and goals of the organization and the efforts of everyone within the network.
- Provide leadership and moral support for the people assigned.
- Ensure that these people receive the education and guidance they need.
- Maintain regular communications within the network for the purpose of sharing experience, learning from each other, venting frustrations, and celebrating successes.

In order to make the network effective, you need to establish, at the outset, a shared purpose and a clear understanding of the nature and the importance of the role. You also need to deal with the mechanics of who does what, when you will meet, how you will communicate, and so on.

You should not find any of this too difficult. You should have some dedicated and talented people involved, and they will help you figure it all out. You should at least ask the president to meet with these people and to reinforce the importance of the task. A better approach is to arrange occasional joint meetings of the Quality Council and the network.

One caution: This group of people has to remember that the improvement process is a means to an end, not a goal in its own right. Like you, they need to understand the organization's goals, keep these in sight, and prioritize their efforts accordingly. Like you, they may need to respond to other priorities in a crisis, although this may impact the objectives of the network.

Line/staff relationships

Your formal relationship with these people may be a "dotted line" one, meaning that they work within another department and report directly to another manager. In some organizations, this mode of operation is common practice and well understood.

However, in most organizations this is not the case. There may be no acceptance of the concept that an individual might receive guidance from more than one manager; or, such arrangements might exist, but function poorly in practice.

Some people may find this arrangement confusing or scary. The individual assigned to the support role may ask "Who is my *real* boss?" Employees may fear being caught between a rock and a hard place—unable to satisfy either manager because of conflicting demands, and subject to pressure and sanctions from both. The onus is on you to get the arrangement clear with your colleagues from the beginning. Then you must ensure that it is explained clearly to those assigned to the role so that they can feel relatively secure and confident.

You should agree with the senior management team at the outset how the dual reporting relationship will work. There is no one right way, but here is a typical arrangement that can be made to work.

- The line manager and the network leader jointly select the support person, agree on this individual's objectives, and agree on his or her performance appraisal. The line manager will usually take the lead in this, with the network leader acting as a contributor.
- The support person continues to report directly to the line manager and accepts this individual's guidance regarding local priorities and the allocation of time to support tasks.
- The support person looks to the network leader for technical support and guidance regarding the change process —that is, approach, methods, tools, and techniques—and to provide a channel of communication with the rest of the network. If the support person ever feels that the guidance coming from the sources is incompatible, he or she needs to know to raise the issue. This is a signal for the two managers to talk. It is the responsibility of the line manager and the network leader to get together and resolve any conflicts, rather than have the support person pulled in two directions.

This is one approach—you may devise a better way of mobilizing and guiding the resources required to support the process.

Developing, nurturing, and sustaining the network

The network develops partly on the job, by trying to accomplish the task in hand, and partly through purposeful action to acquire specific knowledge and experience. Here are some ideas for sustaining the network.

- Establish a regular meeting schedule that does not conflict with the members' other routine commitments to their departments. This may not be easy, but the network cannot function unless the members meet regularly. (Exhibit 7.2 lists typical topics for these meetings.)
- Recognize the need for moral support within the group and empathize with members' problems, but avoid crying sessions.

Exhibit 7.2 *Typical support network meeting topics.*

- Progress against plans, achievements and concerns for everyone present
- Status of some current project or initiative
- Quality Council decisions, directions
- Requests for support from the network
- Nominations for formal recognition
- An educational topic or visit
- Special guests (for example, the president or another senior management team member, or members of a quality-improvement team) to share their experience and perspective
- Feedback and suggestions regarding the operation of the network or network meetings
- Ideas for applying quality principles and tools to the operation of the network
- Input to future plans for the organization, improvement process, and network

Like you, members need to keep team activities focused on the objectives of the organization, and use other ways of venting frustrations. (How people release such repressed feelings is a legitimate subject for discussion, which could be both helpful and fun.)

- Share successes and explicitly recognize each others' achievements. Recognition from within this team may be prized by the individual, but recognition by a member's boss and peers within his or her home department may be more important from a practical point of view. You should, therefore, ensure that individuals receive recognition in their own departments, when this is appropriate.
- Share the work of acquiring knowledge. Each individual might take on the task of becoming the in-house expert in a specific methodology or body of knowledge (process management, statistics, organizational development, and so on). You may find that there is already considerable relevant expertise within the group—you can all benefit immediately simply by beginning to share it.
- Use the internal support network as the link to a much better external information network than any one individual could establish alone. To avoid too much duplication, you may agree to target your efforts in different areas. Some people may be interested in specific professional associations, others may have links with academia.
- Use the network to help with the investigative work required to qualify consultants and other suppliers. Then share information about how these external resources perform.
- Use the network to scout for more talent—each individual should be on the lookout for eventual replacements for him or herself. Members should also be on the lookout for enthusiasts who want to get involved in some way, so that these people can be encouraged and linked with some suitable initiative. As in a volunteer organization, you should strive never to turn down an offer of help.
- Stop occasionally to have fun! Forget the elephant; bring out the pizza!

THE CHANGING ROLE OF STAFF FUNCTIONS

In this section we will discuss the role of staff functions such as human resources, finance, information systems, and quality assurance. We will look at what these departments have in common and how they typically used to operate within a traditional management system. We also will look at

- How their roles need to change in order to achieve their objectives, and to support the organization's objectives
- Strategies for successful transition, focusing on ways in which the Change Agent can help

We will look at the relationship between the internal support network (described in detail earlier in this chapter) and the staff functions.

It is vital that you understand clearly how staff functions can become more effective and valued by changing their roles, because

- Staff functions can be potent allies or lethal opponents of the improvement process. You need to understand the challenges that they face in making the transition so that you can be helpful, or at least sympathetic.
- You may be in the position of managing one of these functions, in which case you will have the task on your hands of leading your people through this transition.
- In leading the support network, you need to ensure that you and others in the network do not start to think and act like a traditional staff function. This is all too easily done—the history of the traditional staff functions is that they all started out as attempts to introduce better practices into organizations, and all succeeded to a large extent. When these functions have deteriorated into dysfunctional mini-empires, this is not because they wanted to, but because they became victims of the system.

What staff functions have in common

Staff functions have much in common in terms of their roles.

- *Staff functions deal mainly with internal, rather than external, customers*—apart from a few activities such as invoicing and accounts payable.
- *Staff functions define their purpose mainly in terms of a relationship with other functions.* This relationship may be viewed (and practiced) mainly as control or as support of these other functions. Line functions sometimes have more colorful terms for what staff functions do, as illustrated in Exhibit 7.3.
- *The work of staff functions is often undervalued.* Because the line functions directly produce something of value to external customers, the potential value of staff functions' contributions is often underestimated. Staff function costs are given names such as *burden* or *overhead* and the functions are often regarded as necessary evils, rather than essential enablers for the line functions.

How staff functions used to operate

Most staff functions used to operate in a fairly similar fashion within the traditional system of management. This mode of operation might not be their choice or their preference, and it might not be the choice or the preference of the line functions, but it is the way that the traditional system typically caused things to happen. The internal competition and adversarial relationships inspired by a traditional approach naturally led to a certain style of operation.

Exhibit 7.3 *Staff functions—support or control?*

Typical quote from a survey of an old-style staff function's internal clients.

"I don't see the point of all these questions about service. It is like asking about a police station using a questionnaire designed for a restaurant. This staff group is just *not in the business* of providing service."

The traditional style of operation looked much like this.

- *The staff function was traditionally considered primarily responsible for some outcome within the organization.* Human resources was considered responsible for proper treatment and compensation of people. Finance was considered responsible for proper financial control. The quality department was considered responsible for the quality of the product.
- *The processes that determine these outcomes are operated by others.* Human resources manages no one but their own staff. Finance does not spend the bulk of the money (and does not know how it should be spent). The quality department does not design the product, put it together, or provide service to customers.
- *The staff functions often attempted to fulfill their responsibilities by interventions that ultimately proved to be counterproductive.* In the process, these well-intentioned efforts often removed from the line functions the responsibility and accountability that properly belong there. These actions sometimes even hampered the line functions' ability to fulfill their responsibilities.

  Finance often imposed cumbersome systems for budgeting and controlling expenditures, and for assigning them to rigid compartments—and, thus, made budgeting a competitive numbers game played between departments. This process guaranteed waste and suboptimization of resources.

  Human resources often imposed on managers various administrative procedures (for example, for conducting performance appraisals), but failed to provide proper advocacy for employees or systems to prevent neglect and mistreatment of employees. In this way, human resources often focused managers' attention on tidy paperwork rather than proper communication with employees and attention to employees' needs and concerns.

  The quality department often inspected the product for defects, rejected defective work, and stopped production when defective output was being produced. Thus, it absolved the line functions of their responsibility to strive for defect-free work in the first place.
- *The staff functions often did not cooperate well with others*—with the line functions or with other staff functions. The staff functions might not cooperate with the line functions because they believed that they were supposed to control these functions—to compel

them to do things right, or at least to prevent them from doing things wrong, (that is, breaking the rules). The staff functions might not cooperate with each other for the same reasons that the line functions might not—because they felt compelled to compete for power and resources.

These situations were not created by the staff functions themselves. They were created by the approach to management that pervaded the organization. They were the inevitable outcome of the prevailing culture and management system.

For example, as long as line functions operate as warring fiefdoms, how can any system be devised to ensure rational budgeting and allocation of resources for the benefit of the whole organization? And, if the budgeting system reinforces these parochial attitudes, how can anyone behave differently? How else can finance attempt to achieve some control over the organization's finances, but to impose a bureaucracy that can be tightly policed by its people, and to intervene when line managers act in a manner that seems fiscally irresponsible?

When there is a lack of leadership skills, people skills, understanding of systems, and so on, then control is out of reach. Costs cannot be controlled properly, product quality cannot be ensured, schedules are often missed, and the management does not know how to make best use of employees' talents. But the illusion of control can be achieved by imposing rules and procedures. Thus, some old ideas, which we now reject, caused staff functions to be misused, and one of their valuable roles—as catalysts for change and improvement—to be lost.

How staff functions are changing

Exhibit 7.4 illustrates the contrast between the way that staff functions used to operate under the traditional system of management, and the way they can operate more effectively. This table demonstrates the magnitude of the transition for which staff functions often need to strive.

Because this also represents a major change in relationships with the line functions, it cannot be achieved unilaterally. The line functions also need to recognize the value of a different relationship, and

Exhibit 7.4 *How staff functions are changing.*

	From	**To**
Role	Customer—for information, evidence, and reports from others	Supplier—of information, expertise and other services
Strategy	Control—by imposition of policies and procedures, and by audit and inspection	Support—by gearing efforts to the needs of others Self-control by client
Goal	Departmental achievement of departmental objectives	Collective achievement of the organization's objectives
Style of working with others	Competitive, adversarial	Integrating, collaborative
Focus of attention	Some aspects of outcomes; for example, product quality, financial results Some pieces of the process; for example, adherence to policy and procedure	The relationship between the entire underlying process and the achievement of all the desired outcomes
Image	Regulator, inspector, policeman	Educator, helper, guide

must be prepared to make adjustments in their approach in order to achieve this.

Staff functions cannot change independently of the rest of the organization, but they can be proactive and lead. They can be influential proponents of change.

Staff functions need to change in the following ways.

- Work to break down the functional silos that traditionally divide the organization. Understanding processes is one of the keys to this.

- Begin the transition from control functions relying upon authority, to support functions providing expertise to help internal clients meet the needs of external customers.
- Progressively transfer responsibility for outcomes back to the people who produce these outcomes—the line functions.
- At the same time, work with the line functions to help them develop the expertise and the effective processes required to produce the desired outcomes.
- Help to implement reliable, auditable measurement systems that ensure that performance in achieving outcomes is visible.

This transition is often accompanied by significant changes in the structure and organization of the staff function. Typically, these changes include transfer of skills, responsibilities, and resources to line functions; decentralization of some operations to get closer to internal clients; and refocusing efforts on work that has more value to the organization. The transition often calls for big changes in the skills profile of people within the function.

As these transitions are embraced, staff functions will find themselves in a more productive and rewarding role—serving internal clients who know what they want, who appreciate it when they get it, and who take responsibility for their own processes. They will also find themselves part of a team that is satisfying external customers, and doing so in an efficient manner—in other words, a winning team.

Difficulties for staff people during the transition

Besides the normal human reluctance to embrace any kind of change, staff people may become concerned or upset about the following issues during the transition of their function.

Loss of importance, authority, and territory

Moving toward a support role doesn't make you less important, but it can feel that way. Being important because others depend upon you for vital information and services feels different from being important because you can make demands and have others leap to your bidding.

Concern about handing over responsibilities

Staff people may feel that it is unwise to trust the line functions because of a perception that they don't care about product quality, proper use and tracking of financial resources, equitable compensation of people, or whatever. Traditionally, the staff function acts as if it is there to force others to do the right thing on these issues. Its experience is that the line functions frequently resist, and the assumption is made that the line functions don't care. In fact, the line functions' attitude is often the consequence of staff function actions.

For example, if you are always trying to stop someone from shipping defective product, you provide yourself as a convenient target for their frustration when the product is defective. Basically, you relieve them of the responsibility of deciding what their own standards are.

Concern about a holistic approach

Many people will not see at first how a holistic approach can work better. For example, people in the quality department may be concerned that broadening the definition of quality to include other issues that affect the customer (for example, schedule, costs, and service) will result in a loss of focus on product quality.

They may have a hard time believing that people who have been trying to slip defective product past them for years do, in fact, care about product quality. They may not see that the operations people have just been caught between a rock and a hard place. Operations people, who are not trained and organized to work on improving the process, end up being repeatedly forced to make decisions about what to do when things have *already* gone wrong—to ship poor-quality product, miss schedules, and/or make costly attempts to recover.

Concern about career prospects and job security

Staff people may be concerned about their career prospects when they see the skill requirements within the organization changing. Some of the most exciting opportunities are associated with the support network, or with the new types of work being undertaken by the staff function in support of the improvement process. These roles may call for additional skills that they do not yet possess.

For example, the quality department people may have invaluable skills related to systems and processes, but may need additional skills related to interpersonal relationships, group dynamics, and facilitation. The human resources people may face a mirror image of the same issue. Here's an opportunity to help each other!

As the staff function moves toward a support role, some of its old responsibilities may be turned over to the line. This translates into reassignment of some people to the line, and/or the reassignment of people to higher value-added work within the staff function, and/or redundancy of some individuals. This is all very unsettling—especially if people feel that they may be working themselves out of a job.

Reactions from other departments

Moving from a control to a support role isn't comfortable when you feel you may have made some enemies in the other functions. Add to this the simple misunderstandings that take place when the basic rules in the organization are being rewritten.

Here are a few examples that relate to the quality department—typical comments that people in this function might overhear from others.

- "After that two-day class, we all understand quality pretty well now. Surely we don't need quality specialists any more."
- "Why didn't the quality department explain all this stuff to us before—don't they understand it?"
- "Inspection is a waste of time. We should stop all inspection immediately and focus exclusively on prevention."
- "Now that quality is part of everyone's job, why do we need a quality department—or a quality manager?"

These kinds of remarks illustrate the truth of the saying "A little knowledge is a dangerous thing."

People in any staff function may be exposed to similar opinions, and may have strong emotional reactions to these. If there has been a history of conflict and adversarial relationships, it may be hard to accept that comments like these are just honest misunderstandings rather than deliberate attacks. Sometimes these *are* deliberate attacks.

Strategies for the Change Agent to support a successful transition

The following are some thoughts on how the Change Agent may help staff functions to develop a more valuable (and more valued) role.

Collaborate with the staff functions

It is vital that the Change Agent work closely with the heads of the other staff functions, to the extent that these individuals are open to this. The staff functions may not seem a priority when it is the line functions that directly deliver the products and services, consume the bulk of resources, and most directly impact the external customer. However, the staff functions are perfectly positioned to act as catalysts and to support the change process—or to undermine it. They should not be overlooked.

Encourage a proactive approach

If staff functions can be encouraged to become proactive in examining the needs of their own internal customers and, hence, reexamining the services they offer and the processes that they manage, then this is a major step forward. The Change Agent may be able to offer encouragement by acting as a role model; for example, by operating his or her own staff function in this way.

Some support processes may be critical, although they are not costly to operate. For example,

- The planning and budgeting process is often a drain on management time and an obstacle to establishing rational decision making regarding the allocation of resources.
- Late, inaccurate, or hard to understand invoicing is often a significant source of customer irritation.
- The performance appraisal process is often a paper chase, rather than a benefit to those who use it.
- The system for reimbursing expenses is often bureaucratic and obstructive, leading to a loss of employee time and goodwill.

Make support functions partners in the support network
Besides getting involved directly in team efforts to improve key business processes, some staff people can play valuable roles in the internal support network—as providers of support such as training and facilitation to other functions, and as champions for change within their own functions.

Encourage others to become role models
Here's a challenge that may appeal to some staff function heads: Influence the line functions by becoming a role model—apply the new approach to your own function. If a staff function will take on this challenge, this is potent support for the process. The line functions will quickly notice the shift in attitudes and style of working, and they will love it. They will wonder why they were not doing this all along.

For example, in a traditional setting, audits often used to be disruptive impositions that inspired fear, and that were to be avoided, delayed, or carefully contained. However, audits can be viewed as a service, and the groups being audited can be treated as internal clients (see Exhibit 7.5). As people learn how to manage their processes and how to use information from audits to help them do so, they come to value audits. (For this to happen, significant changes may also be required in the methods and the attitudes of the audit teams.)

Exhibit 7.5 *Audits as a service.*

As a telecommunications company began the quality journey, the management of one facility invested heavily in understanding and documenting its key processes.

System audits of such operations were planned by the corporate quality function to take place every six months. The management of this facility came to value the information that these audits provided.

When other demands on corporate quality caused these audits to be delayed, the management of this facility complained and became concerned that top management commitment to the improvement process seemed to be flagging.

Help others to develop a plan for the transition

All staff functions will have some defined responsibilities within the organization's plan for improvement; but those which choose to be proactive also need a plan for their own transition, and they may need some help in developing this. In addition to developing new shared goals for the function, this plan will typically involve identifying what they produce, and finding out who uses these outputs and what value the users place on these. This will lead to a reassessment of priorities, identification of the associated processes, and work to improve those processes which are a priority in their internal customers' eyes. All of this needs to be done with the organization's objectives in mind, rather than the narrower objectives that an old-style staff function might sometimes embrace.

During the discussions with other functions, it may become clear that the staff function has some responsibilities that should be transferred to the line. Plans are then required to ensure that line people acquire the appropriate skills, or to transfer staff people to the line function. For example, if the quality department is involved in mass inspection of incoming parts and of product moving through the process, it may decide to transfer these activities to line functions and help reduce the need for inspection by training in defect-prevention methods.

Help the staff function heads with issues that may blindside them

Throughout this process, it is vital for the staff function heads to maintain ongoing dialogue with their people to understand and deal with their natural concerns—for example, about career prospects and job security. Some function heads will have a natural sensitivity to such issues—others may not. Other functions may do a superb job of helping their people through the changes, but struggle with the concepts of systems and processes. They may need special help with technical issues.

Encourage sharing of concerns among colleagues in different staff functions

Because the various staff functions share many common concerns related to the change process, it makes sense to share experiences. This can help build trust and the cooperative relationships that should exist between staff functions, as well as between the line and staff.

Exhibit 7.6 *Staff functions and the support network contrasted.*

	Staff function	**Support network**
Scope	Key subsystems within the management system	The management system as a whole (to avoid subop-timization)
Responsibility	Design and communica-tion of some support processes; for example, for planning and budgeting, information processing, individual objective setting and performance appraisal, compensation, assurance of product quality	Design and communica-tion of processes for improvement Integration of potentially divergent departments and functions
Resources used	People employed by a function (working in col-laboration with people from other line and staff functions)	May ultimately employ few, or no, full-time people May be operated through a team-based structure that draws upon people from all functions May no longer be required at some time in the future—support may be built into the operation of the organization, and/or the improvement process may be considered fully integrated

The internal support network and the staff functions

We described the internal support network in some detail earlier in this chapter. Do we really need such a network? Couldn't one of the staff functions perform this task? Will the support network still be required in the distant future?

This internal support network is almost always required at the outset, because usually no other structure exists that remotely resembles what is needed—a collaborative, nonthreatening, and nonpartisan support system for those embarking on the improvement process. To clarify this point, Exhibit 7.6 lists some of the key differences between the support network and a staff function.

Staff functions—the bottom line

The staff functions are very different in nature from the line functions, and have special issues to deal with during the transition. They may also be powerful supporters of the process, or effective opponents.

If in your role as Change Agent, you also have a staff function to lead, you have your work cut out for you—but, you can make your function a role model for others, and take the lead in cooperating with others. If you do not have a staff function to manage, then you should at least be sympathetic to the challenges faced by your colleagues who do.

8

External Resources

If I have seen further it is by standing upon the shoulders of Giants.

—Sir Isaac Newton

CHAPTER CONTENTS

- The various types of external resources and their different uses
- How to access sources of information and build an information network
- How to select and make proper use of consultants and other suppliers

In this chapter, we will look at the various external resources you can draw upon to help you. Some wise use of external resources is usually essential to the success of a quality improvement effort. Learning how to use them properly will give you the "strength of giants." These resources include

- Sources of information
- Sources of products and services, such as training and consultancy

YOUR INFORMATION NETWORK

There is a wide variety of information sources available to you, including

- Fellow practitioners
- The leading teachers and gurus in this field
- Books, magazines, articles, and other media
- Conferences and seminars
- Professional associations
- Other organizations that are advanced on the quality journey
- Perhaps some of your present customers and suppliers

There are more good sources available than you can possibly use.

Fellow practitioners

When you feel saturated with education and sales propaganda, you probably need some help to sort out what you've heard, separate fact from fiction, and figure out what you can rely on. You need ways of finding your way through a mass of confusing, and often contradictory, information to get the information you need.

One of the most reliable sources of this type of guidance is successful quality practitioners in other organizations. You can make contact with your peers in the field through professional and other associations, by networking at public conferences and seminars,

Exhibit 8.1 *The miracle of networking.*

Sometimes, your network of fellow quality practitioners will allow you to accomplish miracles (or so it may seem).

For example, suppose your president looks up to some other leader whom you know to be a quality enthusiast. The president would love to meet him, but the other person heads up a big outfit and doesn't know your leader from a hole in the ground. You both suspect that your leader couldn't get the time of day from him.

Sometimes you can set up a meeting like this through your network of change agents, in a situation where others have no leverage. Why is this?

- Because your peers in the other organization understand what you are trying to do—and so does their leader.
- Because if their leader is an enthusiast, he will be happy to discuss quality with almost anyone. People often mention this phenomenon—they say "It's like discussing your children."

and through the user groups associated with the leading consultancies (see Exhibit 8.1).

As you make these contacts, you will naturally find out how far along other organizations are in their implementation, and you will begin to sense which of your contacts are respected by others for their knowledge and experience. If you can identify the experienced change agents in organizations that are successfully implementing a quality approach, and establish contact with them, you will get information that is sound and reliable. Given a number of sources like these, you can also gain a variety of perspectives; for example, from different sectors, or in the use of different approaches and methodologies.

Because the people you meet through this networking will usually be helping you as a favor, you cannot impose on them too much. But, they can easily help you to filter all the other information and advice that you may be getting. They will be delighted to help if they can—it's written on the DNA of change agents—and you may begin to build some lasting friendships.

An *unhelpful change agent* is an oxymoron—if these individuals don't seem to want to help, they're probably not the "genuine article" anyway. However, you should be sensitive to their situation. People in their positions, like you, have demanding jobs. In addition, they are often deluged by outside requests for help and information—sometimes by people who haven't even done their homework. Don't be one of those.

These peers can also help you with another vital task—to identify some competent consultants and suppliers. Once you have accomplished this, you have greatly improved the odds in favor of success for your organization.

Over time, you will develop this network into a wide circle of individuals whom you will contact occasionally for specific purposes. Hopefully, this will not be entirely one way, and you can find some way of returning the favor or passing on the favor to others.

Gurus

One of the characteristics of this field is the influence of a few thought leaders and teachers (see Exhibit 8.2)—sometimes called *gurus*—who have made major contributions to the body of knowledge. These people are important sources of ideas and information, but different people approach them in very different ways, so it is worth considering in what capacity you wish to use them.

To illustrate the point, here are two definitions of a *guru.*

1. A guru is someone whose ideas or theories are so powerful that they can illuminate your situation and help you—even from afar, through the filter of media such as books and video. In other words, this person's ideas can help you even if you never employ the individual to come and work with you or to look at the specifics of your organization.
2. A guru is someone whose personal powers seem so awesome, whose insights seem so profound, that others defer to this individual's pronouncements—sometimes out of

Exhibit 8.2 *The vital few: People whose ideas you should study.*

- Dr. W. Edwards Deming
- Dr. Joseph Juran
- Dr. Kaoru Ishikawa
- For others, see Appendix D

> respect more than out of understanding. The novice may believe that the words of the guru must be true because of the source. If the words don't fully make sense yet, the novice will surely come to understand as he or she gains in experience and enlightenment.

This second definition comes uncomfortably close to the definition of a deity. The problem with such an approach to management is that there is no rational basis for deciding what is correct and what is not. Sometimes, people seem to be looking for a guru to follow, and choosing between alternative experts. But why should anyone arbitrarily limit themselves to one source of knowledge?

When you begin to access the teachings of the gurus, you need to decide whether your interest is primarily philosophical, or for practical application to help your organization meet its goals. Theories of management need to be tested. They need to be proven to be valid and useful in the real world. Otherwise, they are no more than dogma that can lead us in wrong directions.

Imagine if students in other fields of science allowed themselves to be led by deities. Picture a guru of chemistry giving out the true periodic table of elements to the world, with the assurance that this version is correct because he or she says so. If chemists operated like this, we wouldn't have nylon or contact adhesive today, let alone carbon fiber and superglue. *Management gurus are not gods, but resources for us to draw on for ideas.* Bearing this in mind, it probably makes sense for us to

- Become familiar with the ideas of many teachers—not just one guru—and take from each what we find illuminating and helpful.

- Test these ideas against our own experience, and fully accept only what we fully understand—knowing that our personal experience and understanding will grow in time.
- Take on trust, for the moment, what seems to make sense—but test it in practice and find out what other experienced people working in the field think.
- Look for evidence from the field regarding how well these theories hold up in practice.

Books, magazines, and articles

If you got this far, you belong to a rare species—people who not only buy books, but read them all the way through. In terms of volume of words read, or information received, most people get the bulk of their input about business issues from magazines, newspapers, and television, not from books.

Bear this in mind with your colleagues. As someone with a special interest, you may put in the time and effort to acquire and to study books on quality. But when did you last read a book on accounting or marketing, or any specialty other than your own?

Your colleagues are more likely to get their information about quality in snippets from business papers and magazines. This is how they also become targets of the media. The opportunity for you is to capture relevant and accurate information and interpret it for them.

Communicating the message

Messages carried by the media will inevitably be an issue for you in your work, because these create noise in the organization's communications system. Whenever a newspaper or magazine publishes something that seems related to your role, you can be sure that some colleagues will cut it out and send it to you. More than that—they will expect you to have read the article already.

Every week another article is published that tears down as folly the wisdom of last month's articles. Every month a new management book emerges that claims to capture the essence of human wisdom, compressed into five platitudes and formulas that can be put into practice in ten minutes. How can you deal with what may begin to feel like a barrage of distracting fluff and misleading propaganda?

The way in which you go about spreading the word will determine the vulnerability of your message. Here are some ideas.

- *Minimize the use of jargon.* Terms like *total quality management* or *TQM* are helpful shorthand for practitioners to use among themselves, but they are definitely jargon. Even the word *quality* is jargon. This is demonstrated by the fact that quality practitioners are trained to use the word in ways that the rest of the population don't understand.

  There is nothing that can be said using such jargon that cannot be said equally well using unambiguous everyday language that everyone can understand. At worst, this may require a few more words.

  One way of reducing the use of offensive jargon is to make up your own in-house terms. For example, your team may create a name for the improvement initiative in your organization. This provides the team with another opportunity to take ownership—and avoids the use of outsiders' jargon.
- *Avoid dogma.* We have already spoken about the danger of helpful teachings and ideas becoming the domain of quasi-religious fanatics. This attitude is likely to antagonize others rather than persuade them, and there is no need to adopt such an approach.

  The basic principles of quality are unassailable: Customers are more likely to come back if you satisfy their needs; people will do a better job if they know what's required and have the knowledge and tools to do the job; prevention is better than cure. Who can argue with such ideas? However, if your approach is to recite "Dr. Deming says this," or "Dr. Juran says that," then to many people you might as well be quoting Dr. Spock.
- *Don't try to defend the whole universe of truth—just the specifics that are important to your process.* For example, suppose a colleague saw a well-written article in a reputable journal, claiming that "TQM is now passé." What should your response be?

  You might indicate that this is very interesting, and that you would like to see the article and study this new information. But bring the conversation back from generalities to the specifics of what *your* organization is doing. Does your colleague feel it's no longer a good idea to find out what our customers think of us? Of course not!

• *Take the initiative in sharing relevant articles and information.* When articles contain apparently negative information you can often use this in a positive way; for example, by putting the information in context, by pointing out lessons to be learned, or by qualifying the source. If, however, they offer solid grounds for questioning what you are currently doing in your organization, you should investigate.

• *Do not dismiss negative or seemingly trivial information without examination.* You might learn something.

- Too many organizations do get stuck or fall by the wayside on the quality journey, and someone should be doing sound research into the reasons why.
- There are many excellent articles published that crisply and accurately summarize new developments.
- Some whimsical, easy-to-read publications are accurate, and even contain "nuggets" of insight.

You need to keep abreast of the latest ideas and publications.

Professional associations

The American Society for Quality Control (ASQC) is an excellent source of information for quality practitioners. Articles in the ASQC's monthly journal, *Quality Progress*, are peer-reviewed to ensure accuracy, as are ASQC books, such as this one. The ASQC book catalog is a gold mine, listing many of the best publications available in this field, regardless of the publisher. Local ASQC sections provide networking and educational opportunities, and opportunities to make a contribution, as well as to benefit from contact with peers.

There are other associations that cater specifically to quality practitioners, and many sectorial or professional associations offer quality-related information and education. The American Production and Inventory Control Society (APICS) is an example of a specialized professional association that has been very active in this field.

Public conferences and seminars

Public conferences and seminars can be invaluable for networking, but you could spend your entire working life attending conferences and seminars. Here are some ideas on how to select those that will help you.

- Don't just sift through the pile of brochures that came in the mail, looking for something interesting. That's not education, that's tourism. Choose conferences or seminars based upon your learning needs. Start with a clear understanding of what you want to find out, then seek out suitable opportunities.
- Use your network to identify possibilities and then to qualify likely events. Who is the organizer? Is this a regular event? What were previous ones like? These are the questions you need to answer before you invest your valuable time.
- Adopt an organized approach within the senior management team and within the internal support network. Suggest that one person go to any promising conference (chosen according to needs and interests) and that when he or she gets back, a presentation is made to the others to summarize what was learned.

Awards administrative bodies

The bodies that administer the various national awards programs publish some extremely useful material, including

- Information about past award winners and how to contact these organizations.
- Case studies and videos outlining the practices used by high-performing organizations.
- The awards program criteria. These are invaluable instruments for self-assessment—as well as for other educational purposes.

In some cases, training materials, such as case studies used for training the award examiners, are also published. Appendix D lists some of these organizations.

Other organizations implementing a quality approach

Local organizations that are on the quality journey can be identified through your network of local contacts. Organizations that are at the forefront in quality can be identified by finding out who has won the various national and regional awards in North America, Europe, and Japan. There are also a few long-standing supplier awards programs operated by large organizations such as NASA, and you can find out which companies have won these awards. The companies that win such awards are often well-organized to share information about what they do, in the form of reports and sometimes through visitor programs.

For most quality awards, one of the requirements is to demonstrate that the organization is already sharing this knowledge with others in a proactive way. The award is unlikely to be won by a company that is not already doing this.

If you do choose to use one of these companies as a resource,

- Do your homework first. If you use a master only to help you learn the basics you will gain less from your contact with them and you will have used up some of their goodwill for little purpose.
- Focus your attention on the specific division or facility that won an award. Regardless of corporate claims to the contrary, the winning of an award by one division or facility often tells you little or nothing about the capability of the rest of the organization.
- Draw up a short list of issues you want to investigate. Few organizations do everything well, and there's no sense in exploring areas where they do not do a great job—or which are not relevant to the areas you are trying to improve.

Existing customers and suppliers

Customers are always the primary source of information about what aspect of your products and services need to improve—and whether your efforts to improve are working. Suppliers can help you understand how to make better use of their products or their particular technological expertise. However, for information about quality improvement *methods*, customers and suppliers are often no help at all.

If you are really fortunate, you may find that one of your existing key customers or suppliers is an organization well advanced on the quality journey. In this case, you have probably struck gold. This discovery opens up all kinds of possibilities for working closely together, because you already have a relationship and a reason to help each other.

CONSULTANTS AND OTHER SUPPLIERS

External help of some sort is usually essential in order to accomplish major change, and there are outstanding resources—people, services, and products—available to help you. In this section we will discuss how to find and make best use of these. These suppliers include

- Consultants
- Training companies
- Survey companies

Reasons for using external suppliers

There are very good reasons for using external suppliers. *In fact, most organizations cannot hope to be successful without making some judicious use of outsiders.* The following are typical reasons why you should decide to use outsiders.

- You need to draw upon expertise that does not exist inside the company.
- You need temporary additional resources for some project.
- You need a task performed that demands the independence and objectivity of an outsider, or the additional credibility that an outside expert is accorded.
- You need materials that would be costly or impractical to develop in-house.

For example,

- You may want some advice from one of the world's leading experts on benchmarking. Or, you may simply want to draw upon someone who has some recent hands-on implementation experience.
- You may require a specialist survey company to gather information from your customers or employees.
- You may need interpersonal skills training materials (and instructor certification classes) from a training supplier.
- You may require an external quality assessment, and the feedback from this, delivered to the top management group by a credible (and fearless) outsider.
- You may want an experienced facilitator/trainer to support early problem-solving or process-improvement teams until your in-house support people get up to speed.

These outsiders are not the only sources of information and guidance. In fact, although you pay for their services, they are not necessarily the best sources for every type of information.

At any given time, many of the most experienced and knowledgeable practitioners in this field are not working as consultants or suppliers. Many of these people, who learned their craft implementing a quality approach within their organizations, are still working like you, as internal change agents.

Although these people are kept busy working for their own organizations, you can get help from them, as we already described. Even a little input from them may be of great value to you. These individuals have the great advantage, from your perspective, that they are not trying to sell you anything.

Working with suppliers

If every potential supplier—consultancy, training company, or survey outfit—were a quality organization, your life might be simpler. Regrettably, some are not. The suppliers competing for your business may range from outstandingly competent and professional, to incompetent and unscrupulous. Poor decisions regarding the procurement of outside support can cause you big problems. In order to safeguard your interests, this section errs on the side of caution. The watchword here is *caveat emptor*—let the buyer beware.

There are many first-rate suppliers in the quality arena, and the tone of this section is unfair to them—they are exemplars, who do not deserve to be tarred with the brush of dubious practices, or treated with suspicion. However, these suppliers will welcome and stand up to the scrutiny suggested here. The intent of this section is to benefit you, and them, by helping you to identify the most competent and suitable suppliers available.

When working with suppliers for quality-related products and services, you will be purchasing

- The knowledge, experience, and skills of an individual
- A product, such as training materials
- Intellectual property, such as methodology
- Some combination of these

The mix of these elements involved in any given situation influences how you select a suitable supplier. In some cases, you need to focus on finding out about the individual(s) proposed for the task, while in other cases it is mainly the product or the approach that you need to evaluate.

Understanding the supplier's perspective

It usually helps to understand how suppliers make their profits, because that determines what they are motivated to sell you, and, therefore, where their bias may lie. For example, a supplier may make its profits mainly from selling training materials, ranging from instructors' manuals, overheads and videos, to workbooks for students. All other offerings from such a company—such as introductory seminars, executive education sessions, certification of in-house instructors—are likely to be viewed primarily as a means of supporting the sale of more training materials.

What does this mean for you as a potential customer? It means, for example, that the executive education session, that you view as an important *high-value service* vital to your start-up process, may be viewed by the supplier as part of its *sales process.* You may not be purchasing education and guidance, but a sales pitch.

This does not mean that you should not use such introductory services. The executive session may be excellent. You may use it to accomplish your purpose, and it may be worth every penny. However, you should be clear up front what you are buying. If you procure executive workshops from a company whose income comes mainly from training, you may be confident that the executive team will come away convinced that this supplier's particular type of training is a vital part of the solution—if not the whole solution.

Finally, you should never simply accept the way the supplier organization categorizes itself. It is your right (and your responsibility) when you are qualifying suppliers, to determine what business they are in—their primary areas of expertise, and their primary sources of revenue.

Consultants

Consulting is potentially an extremely valuable service, and well-executed consulting work can help you beyond measure. But a consulting assignment can go wrong in any organization, and when this happens it is the client who suffers. The consultants can suffer too—they will undoubtedly feel bad for a while, and the incident might hurt their business if word gets out. But the client has to live with the consequences of the disaster—perhaps for many years.

A frequent cause of problems is that clients put too much faith in consultants, or do not know how to select them and work with them. The following sections are designed to help you gain access to the valuable outside resources and expertise you may need, without exposing yourself to unnecessary risk or cost.

How consultancies operate

As a business, consulting is dominated by the fact that the product has no shelf life—a consultant's time cannot be put into inventory to be sold later. This means that there is intense pressure to keep consultants on the clock; that is, spending their time on billable work. This is not necessarily value-added work from the client's point of view.

Such pressure creates a strong short-term focus, and a need for consultants to be alert to trends. If there is a fall-off in demand for *X*, perhaps the consultants can keep busy by going after the emerging market for *Y*. This could also be called following fashion, and this practice can sometimes work against the client's need for in-depth knowledge and implementation experience within a specialty.

Consulting is primarily an individual activity. There are many exceptions, but most consulting is delivered by just one or two individuals, and the job generally requires a fair degree of self-reliance. Because of this, consultants tend to operate in a fairly independent fashion, forming alliances of various sorts where this is convenient. Independent consultants often form networks of associates for marketing purposes and in order to combine forces on contracts that require many people.

Large consultancy firms often extend this concept rather like a franchise. Each small unit or division is essentially an independent business. The consultants receive leads, office space, administrative support and use of the prestigious logo—and the consulting firm receives a proportion of their billings.

One of the major challenges for consulting groups —large or small—is to achieve effective internal cooperation, so that jobs are proposed and operated by the most suitable and well-qualified consultants, rather than by those who happen to get the lead.

For the convenience and other perceived benefits of using a major firm, the client pays a premium. The per diem fee rates for the same individual may be 50 to 100 percent higher if the consultant is working for a large firm than if the same individual is operating as an independent.

Selecting consultants

In this section, we will discuss do's and don'ts in selecting consultants (see Exhibit 8.3).

- Focus on the experience of the individual(s) who will do the work, not the capability of the consulting group.

There are sometimes advantages to using a consulting group that has substantial resources and a reputation to preserve. For example, some colleagues may perceive the consultant as being more credible because of the well-known logo on his or her business

Exhibit 8.3 *Dos and don'ts in selecting consultants.*

Do	Don't
• Write down the objectives and deliverables, and get these clear. • Consider alternative ways of getting the job done. • Set out the criteria for selecting the consultant. • Focus on the relevant experience of the individual(s) who will do the work, not the group or the salesperson. • Always ask for references for similar work, and always investigate these references. • Look for hands-on implementation experience. • Ask the consultant to explain the limitations of the approach proposed. • Assess the potential working relationship, as well as technical expertise.	• Don't be overly impressed by – First impressions, demeanor, or presentation style (at the expense of substance) – Paper credentials – Authorship of a book that you haven't read – A prestigious address or plush premises – The appearance of great success or wealth – A big-name client list (Over time, large organizations may engage *thousands* of consultants.) • Don't set utopian standards, beyond the real needs of the task.

card. Or, the scale of the task may call for a large team of consultants, for example, to operate on a national or international scale.

However, for many purposes the capabilities of the group as a whole are irrelevant, because the work will be done by one or two specific individuals. It is their expertise you are paying for, not the capability of someone else whom you may never even see.

A standard practice is to put in a proposal the résumés of the best-qualified people as being typical of what the group can provide, although others will probably be offered to do the work after the contract is sold. The implication of the proposal is that the group can choose from an army of superbly qualified people. In reality, even in a large operation, there may sometimes be only one or two people who are well qualified to meet your needs. But when the job is sold, you will be offered only people who are available when the work is to be done. Also, because of the franchise-like nature of many such firms, the division that received the lead may submit the proposal and try to do the work with its own people—unless the task is clearly beyond their capabilities, in which case, they will pass the lead on to another division.

Another possibility is that the proposal is sold by an impressive (and truly expert) consultant, who inspires confidence and passes every test—but who is rarely seen again once the contract is signed. Again, focus on knowing the capabilities of the individuals who will actually do the work.

- Always ask for and investigate references for similar work.

You will usually use references to qualify specific individuals. However, you will also use references in order to check out a training package or a methodology proposed. Consultancies do not readily hand out extensive lists of their client contacts, but they should be prepared to put you in touch with a few references. You may ask for a list of other organizations for whom similar work has been done, and then ask for contact names and phone numbers for a few that you choose.

Then investigate these references! It does not matter how impressive or clear-cut the story seems on paper. The client doubled output in 12 months? Perhaps the client considers that they accomplished this *in spite of* the consultant.

Be sure to find out from your contact which consultants did the work and the client's view of these people. Also, find out whether the work is complete and the final outcome satisfactory to the client. The contact is almost useless as a reference if he or she is still in the honeymoon stage—following the plan in earnest hope and expectation of a satisfactory future outcome.

Another reason for checking references is that consultants routinely rewrite their résumés to bring out experience relevant to the job and to remove irrelevant information. This is legitimate—you probably did the same the last time you were applying for a job. However, you should look for specifics. Being "involved from the start" in a previous similar project could be code for "helped write the proposal."

- Look for implementation experience.

Because the *doing* of quality is so much harder than the *understanding* of it, hands on implementation experience is the main training ground for helping others to implement a quality approach. This is like riding a bicycle, playing golf, or flying a plane. How else can one learn? By consulting? By observing? Only by doing!

Exhibit 8.4 *Some characteristics of good consultants.*

Good consultants—those who can really help you—typically have many of these characteristics.

- They are good at building relationships with others.
- They are expert in the consulting process, including interpersonal communications, interviewing, gathering and analyzing data, giving presentations, report writing, and so on.
- They have subject-matter expertise in some specific areas. This is usually the core of what you are paying for, and, in quality, this usually means solid implementation experience.
- They have a system perspective—a way of seeing the big picture and understanding the underlying dynamics of the situation.
- They are good at relating their knowledge (from other situations) to your specific situation.

People who lack practical experience may be outstanding at some tasks: conducting research, analyzing research results and presenting the findings, articulating the concepts of quality in an interesting way, and raising interest and awareness. However, when you encounter issues that demand the insight of having been there, they will be unable to offer reliable advice. They may make an educated guess, or repeat what they have heard, or fall back on a research finding—but by relying on this guidance, you could be flying blind. Would you take flying lessons from someone who had never handled the controls, but had only been a passenger?

- Ask consultants to define the limitations of what they offer—what they can and cannot do, and under what circumstances their services would be counterproductive. Every pharmaceutical product on the market has listed specific circumstances in which the drug may help and circumstances under which the drug should not be used, because it may do harm. Similarly, every consultant's

Exhibit 8.5 *Some suggested questions for consultants.*

Methodology
- When is this approach particularly appropriate?
- When would it be a bad idea to use this approach?
- What alternative approaches exist?

Consultant capability
- What types of work do you specialize in?
- In what type of work do you have a lot of experience?
- What type of work would you not attempt?
- What assignment were you working on last week? (Was this related to quality improvement?)

References
- At what other organizations have you applied this approach?
- Which of these organizations has completed its implementation (and can judge effectiveness)?
- What kinds of difficulties have your other clients encountered with this approach?

knowledge has boundaries, and every product or service that you might procure has its strengths and limitations. If your suppliers cannot define their limitations in a way that you can understand, then either they do not know, or they are not being candid, or both.

• Assess the potential working relationship, as well as technical expertise. When you employ a consultant you are beginning a relationship that may be important to you now and that might develop into one that is long-term. It makes sense to seek out individuals whose values and style of working fit with yours and with what you are trying to accomplish. Perhaps the most important thing to ask yourself is "How far can I trust this person to act in my best interests and the best interests of my organization?" This is an important issue and one to explore when investigating references.

• Don't set utopian standards. The thorough approach suggested here may reveal weaknesses, but the task you have in mind may not require the world's greatest expert. A conscientious and trustworthy consultant with the basic competencies required may do a fine job for you. Exhibit 8.5 suggests some of the questions you may ask when selecting a consultant.

Working with consultants

The following are some suggestions to consider when you begin to employ a consultant.

• Agree on some ground rules beforehand with your colleagues. As the organization's designated in-house expert, you should have the final call regarding any consultants used for quality-related work with the top management team. Consultants whom your colleagues use within their own divisions are a different matter, but there is a need to strive for some consistency of methodology and language throughout the organization. You will probably want a team discussion about how to avoid having a dozen different horses hitched to the wagon—all pulling in different directions.

• Having gotten clear, documented assignment objectives agreed to at the outset, make these the basis for the contract with the consultant.

• Set ground rules with the consultant. As the Change Agent, you are usually the client for consulting work related to the change process, not the president or your colleagues. In this case, the consultant is there under your supervision to help the team execute the plan, not to pitch his or her ideas over your head.

An experienced quality consultant will naturally adopt the kind of support role set out here and will not even need to be told. But, you need to be sure that he or she does understand what is required.

Having undertaken all of this careful scrutiny and evaluation, you should have set the stage for a satisfactory consulting assignment—and perhaps for a mutually beneficial long-term relationship.

When the assignment is complete and you are paying the invoice, don't forget to offer candid feedback to the consultant. Like most people, conscientious consultants thrive on a sense of having accomplished something useful, and on sincere appreciation of their efforts. Good consultants are open to suggestions for improvement, and take these very seriously. They need the check to live, but they may value your appreciation and constructive feedback just as much.

9

Personal Survival and Growth

But of the good leader, when his work is done, his aim fulfilled, the people will say "We did this ourselves."

—Lao-Tsu

CHAPTER CONTENTS

- The personal challenges and characteristics of the job
- The experiences of others in this work, and their advice—based on interviews
- Looking after yourself
- Knowing yourself and making a personal decision about doing this type of work

Why do people do this type of work? What are the rewards? What are the drawbacks? Do you really want to take on this type of task? This chapter tries to help you answer such questions. It deals with what it feels like to work in the role of a change agent, and the kinds of personal challenges that arise. It provides ideas on how to rise to these challenges, and how to look after yourself, professionally and personally. It summarizes a series of interviews in which practicing change agents reveal the experiences that were significant to them—both high spots and low points. Finally, it deals with knowing yourself. This is the most important issue in discovering whether this work is for you.

WHAT THE JOB IS LIKE

The people who seek out this kind of challenge tend to be plucky and resilient. But, without information about the nature of the job, they may not only fail their colleagues, but neglect their own needs and suffer needless pain and anxiety. Even worse, people may go into this job for the wrong reasons, and find themselves unhappy and ineffective because the job is not what they expected.

Here are some of the characteristics of the job of a change agent. This job is very much like parenthood in that

- It is not for everyone.
- It is very rewarding when it's going well and correspondingly awful when it's not.
- You acquire a whole new range of skills and discover new opportunities.
- It helps to build character and maturity.
- The main rewards are nonfinancial.
- The outcome is not completely under your control.

It is a vocation

This job is a vocation because

- You need to believe in it to do it well.
- It offers rare opportunities to make a difference.
- It is short on conventional rewards and recognition.

You need to believe in it

There are many occupations that people cannot perform well unless they want to do this type of work and they strongly believe in what they are trying to accomplish. Being a change agent is one, because

- You will not succeed in influencing others and sustaining their enthusiasm unless you strongly believe in the value and the legitimacy of this change process.
- You will not be able to find the energy, commitment, and persistence required to keep at it yourself unless this is something you really want to do.
- You will not become a credible role model for the style and behaviors you are promoting unless the underlying values are part of your personal makeup.

It offers rare opportunities to make a difference

This work has the potential to affect greatly the lives of many people for the better and to help many others to grow personally. It is hard to describe the enthusiasm and pride that radiate from employees who are accomplishing things they used to view as impossible, or who have learned to work together and support and value each other.

Few people work for purely altruistic motives, but most of us would like to feel good about our work—to feel it is important or valuable to someone. There is great satisfaction in seeing an organization change for the better, experiencing the effects of this on everyone involved, and knowing that you have helped to make this happen.

It is short on conventional rewards and recognition

The rewards are indeed great, but they are not the conventional ones. Picture this situation:

> *The accounting department has achieved unheard of levels of speed and accuracy in the billing process. Customers have quit cursing the company's awful invoices. Even salespeople, in unguarded moments, have been known to comment on the improvements. The team that accomplished all of this is being recognized at a special meal and presentation ceremony. You, the Change Agent, are taking part as one of the senior executive contingent.*

As the speeches and the toasts go by, you reflect on the past year. You recall the cajoling and negotiating you had to do just to persuade the vice president of finance to agree to allow such a team to be formed. You think of how you have run interference for this team ever since, protecting it from being dissolved when it became evident that the team was getting somewhere (and would reveal some embarrassing shortcomings). You think of the hours you spent with the team leader, persuading him of the importance of paying attention to the feelings of team, which was coming close to mutiny over his high-handed style.

You decide that you had quite a lot to do with this accomplishment today. You feel pleased with what has been accomplished—and with your own contribution.

Great! Because that is all the reward you should expect.

The people directly involved, from the vice president of finance to the team leader and team members, all consider this their accomplishment. They provided the resources, or performed the actual work, and they all struggled at times to meet their commitments to the project. If they are to look back on this as a positive experience and want to repeat this effort, they need to be recognized and encouraged.

In fact, it is a vital part of your job as a change agent to ensure that the participants receive recognition for their efforts. It is all too easy for management to simply accept the achievement and switch its attention to the next challenge. So you prime the president, remind the vice president of finance, and coax the recognition coordinator into action. That's your job. If someone—the president, the vice president of finance, or even the team leader—remembers to thank you or acknowledge *your* contribution, that's a nice bonus.

It is an opportunity for personal and career development

The change agent who has some experience under his or her belt has developed a deeper insight into human nature and a greater facility for working with others. He or she has also honed a broad range of general-purpose management skills, from strategic planning and project management to group facilitation.

And, of course, the change agent has become proficient in various quality-related methods and techniques which are increasingly sought after in leading organizations—from process improvement and benchmarking to team-based approaches to getting work done. An individual who has such skills and experience will always be sought after.

The job also offers a uniquely broad scope and variety. Few jobs, short of the president's, offer such a broad perspective of the organization, or the opportunity to get involved with every department and location. If you enjoy being able to see the big picture, rather than being confined within one department, you will enjoy this experience and learn a great deal about how organizations work.

It is personally taxing

Jobs that offer great satisfaction and opportunities are usually demanding, and this role is no exception. Here are some of the ways in which being a change agent is a personal challenge.

It involves big highs and lows

No matter how well-conceived the plan, change always takes place in a somewhat chaotic fashion, and breakthroughs and setbacks both occur at unpredictable times. For example,

- On Monday, the first process improvement team, which you have personally championed, coached, and protected, gives its recommendations to the new Quality Council. Contrary to your fears, the team's recommendations are well received, and they are treated courteously. What a nice surprise!
- On Tuesday, your most enthusiastic team member announces her departure. This is a line manager who is a rising star and who has really caught the quality bug . She got a job offer elsewhere that is too good to refuse. Disaster!
- On Wednesday. . . .

The surprises may not come quite so thick and fast, but orchestrating change often feels like being on a roller coaster—soaring like an eagle one moment, and then plunging the next like a plucked turkey.

It involves pursuing goals that are often intangible

Here is a typical comment from a recently appointed Change Agent: "I have trouble motivating myself in this new job, because I find it hard to identify concrete goals and accomplishments for myself." This is a telling remark, because some of the most important goals that a change agent pursues are intangible in the sense that they are related to behaviors and attitudes rather than the accomplishment of output goals or completion of project milestones.

Here are typical landmarks—intangible events that might be signs of important accomplishments.

- When a major problem was revealed during a project review, the president asked for information and analysis about the causes, rather than reprimanding the messenger on the spot.
- The vice president of marketing volunteered the view that many product-related customer complaints were not due to manufacturing defects, but were caused by salespeople exaggerating the capabilities of the product.
- The members of a key problem-solving team realized that their approach had been focused on symptoms, and redirected their efforts to focus on root causes.

The message is: you have to learn to gauge your accomplishments in different ways—and to celebrate and feel good about breakthroughs that don't mean a thing to the casual observer.

It can be lonely

There are various reasons why the Change Agent may at times feel isolated, unappreciated, and vulnerable.

- There are times when no one—including the president—gives a hoot about improving quality, because the world is caving in from some other direction. This crisis has to be dealt with before anyone can focus on anything longer-term.
- People working toward very tangible goals—like 2000 policies sold this month—may feel that the Change Agent has an easy job. They may think so, and they may also say so: "It must be great having no stress and no targets to meet," or "You're a capable person—why don't you get yourself a better job?"

- When significant change does begin to take place, some people will be quite upset at times, and the Change Agent can become a lightning rod for their reactions. In their eyes, he or she becomes a threat, or a symbol of what they don't like about the new order. No one enjoys being on the receiving end of such reactions, but the Change Agent may not understand what is going on. Perhaps this individual takes abuse and hostility personally, and will begin to wonder what he or she is doing wrong or what is wrong with him or her. Even when one does understand that it's not personal, it is still no fun. Social workers receive training to enable them to understand and cope with such reactions from their clients. Change Agents sometimes don't even know to expect this.
- The Change Agent may not have enough contact with others who are in similar roles. Sometimes this individual has little idea of the extent of the quality revolution, and begins to feel that he or she is the only person on the planet in this type of situation. Making contact with peers in other organizations, and comparing notes, is like an injection of adrenaline—it reaffirms the validity of what the Change Agents are attempting to do and provides answers to some problems. It reassures Change Agents that they are not alone, and that they are not going crazy. It restores confidence and enthusiasm.

It can be risky

In terms of physical danger, being a change agent is a lot less risky than being a convenience store clerk or a cop in Los Angeles. This occupation carries other risks—to one's career.

Organizational change takes place in a somewhat chaotic fashion, and so accidents take place. Just as in every war, some troops are accidentally shot by their own side, so those who are on the front line are more likely to collect stray bullets than those who are keeping their heads down in the rear.

As an example, picture the following situation.

The president feels fully committed and believes that she is supporting the initiative. In reality, she keeps falling back into old habits and sending the wrong signals. This situation is obvious to others and is beginning to undermine the enthusiasm of those who are working hard to improve. Others are losing heart and questioning the president's commitment.

What should the Change Agent do? Confront the president directly? Approach the topic obliquely? Wait three months for the next feedback survey to deliver the message?

A Change Agent will inevitably encounter some situations where there is no great move, and where the right thing to do might prove to be career-limiting. If you want a quiet life with few risks or surprises, this isn't it.

It is never complete

Unlike managing a project with a completion date, or working toward well-defined monthly targets, there are fewer occasions when a Change Agent can say "We've met our goals—we can relax for a moment and celebrate." Of course, there will be quality-related projects with clear milestones and tangible successes to celebrate, and these are important. But, the underlying task of changing attitudes and behaviors is never complete. One can never say "I've done everything I should have done."

REAL-LIFE EXPERIENCE—THE AGONY AND THE ECSTASY

This book is based upon my own experience, but it is also the distillation of the experience of many practicing change agents. Every chapter has been carefully reviewed by a number of other experienced individuals working in this field. More than 30 people helped me by reviewing parts of the manuscript at various stages. The aim was to ensure that the ideas and the guidance expressed here were more reliable than just one person's opinion.

This chapter goes a step further by asking some of these people to tell us in their own words about their experience. The purpose is to bring to life the day-to-day realities of this type of work.

For the material on the following pages, I conducted a series of structured interviews with a sample of other change agents, for the specific purpose of probing their experience in more depth. The people chosen for these interviews ranged from seasoned veterans

Exhibit 9.1 *Change agents interviewed.*

Doug Bell, senior director, Petro Canada Resources, Calgary, Alberta; David Carlson, corporate coordinator of service quality, Transport Canada, Ottawa, Ontario; Lynn Cook, vice president, support services, University of Alberta Hospitals, Edmonton, Alberta; Bill Delroy, assistant vice president, Bell-Northern Research, Kanata, Ontario; Roger Gagnon, director of quality, Bell Canada, Hull, Quebec; Frances Horibé, fundamental review consultant, Transport Canada, Ottawa, Ontario; John Long, senior principal consultant, Ernst & Young, Toronto, Ontario; Karin Lunau, manager, organization effectiveness, Worker's Compensation Board of BC, Richmond, British Columbia; Peter McCulloch, quality training manager, Xerox Canada, North York, Ontario; Bob McGrath, president, Quality Quest, Pittsburgh, Pennsylvania; Jim Ludtke, director, quality assurance, NCR Canada, Waterloo, Ontario; Bob Fisher, manager, continuous improvement, Consumers Glass, Etobicoke, Ontario.

in this type of work, with considerable successes under their belts, to relative novices in this field, with the learning experiences of their first few years still fresh in their minds. (These people were already experienced managers before they took on this type of task.) All are people whose views I respect and value. (See Exhibit 9.1 for a list of people interviewed and Exhibit 9.2 for interview content.)

I believe that this small sample of people is fairly typical of change agents across North America, and that the experiences they reflect in their comments are also typical. Where the comments offered seem very personal, I have changed some details so that the individual cannot be identified. Here is what they had to say.

Becoming a change agent

"I was drawn to this type of work by the variety (there's always a new area to deal with), by the fact that I'm always learning, and by the constant challenge. There's never a dull moment, and I'm always on the move."

Exhibit 9.2 *Interview content.*

- What led you to get into this type of work—what attracted you?
- What kinds of surprises did you encounter as you started the job?
- What was the hardest thing for you to learn or to come to grips with (from a task viewpoint, or emotionally)?
- What were some of the most personally rewarding events or incidents as a change agent?
- What types of situations or incidents caused you the most personal struggle?
- What do you feel was the best or most significant decision you ever had to make in the job?
- With hindsight, have you made any decisions you regret?
- What accomplishment are you most proud of?
- What is the one piece of advice you would offer someone starting into this type of work?
- How do you deal with the stresses of the job?
- Is there anything else that change agents should know about their personal survival and growth?

"I was already an experienced line manager who had seen others orchestrate change. I became convinced that there was a need for change in our organization, I felt I knew what needed to happen, and I figured I could do it. I went into this with my eyes open."

"When I was in high school I knew that I wanted to help people be productive and happy in their work. This led to career guidance and then to organizational development work. I believe that people want to make a meaningful contribution—to be productive."

"I decided that my job (vice president of operations) did not provide enough scope to satisfy my need to make a difference. I had full control over my large department, but I could only make a difference in a wider sphere by encouraging many others. So I asked the president to allow me to take on a new, additional role of leading the development of the whole organization."

"I got hooked on employee involvement in 1982, when first I saw the power of applying a group problem-solving process. I saw this transform the working of groups which had previously been a complete rabble."

"I was at a stage where I wanted to broaden my horizons and do something different, but I didn't know what this might be. My boss invited me to take on this new job and gave me a book on Deming. When I had read this, my first reaction was 'There's no way we'll ever do this stuff in our organization!' We talked a lot and I became attracted to the challenge. I thought 'This might be fun!'"

"I always had great empathy for customers. I like to be treated properly as a customer, to get what I've been promised, so I have strong feelings about treating our customers right."

"In 1966 I was introduced to the ideas of Deming and Juran by my manager, who became my mentor. No one had heard of these ideas then, and it was an uphill struggle. As I became more senior as a line manager, I was able to implement a quality approach in whatever department I was running. That was the only way I could do it. Now I operate as a change agent from a line management position."

"I didn't have much choice. I was an operations manager when two divisions were amalgamated and many positions eliminated. I drew what I felt at the time was the short straw—I was given the quality department. However, I decided to make the best of it—to create a quality department 'like the one I always wanted.' I've never looked back."

Initial surprises

"I was surprised at how people would keep saying 'Why do we need to change?' even though it all made perfect sense to me."

"I could not inspire people to change as easily as I had thought. I was amazed at how some people are welded to the status quo by feelings like 'This is what I've built for myself, this is where I belong'—even though they are not satisfied with the present situation and complain endlessly about it."

"I was pleasantly surprised at how much enthusiasm I found—I hadn't expected this. I discovered that most people *were* quality conscious. Usually they wanted to do the right thing, but were just misguided."

"Some of the strongest supporters were crusty first-line managers who appeared at first to be opposed. Of course, some of the crusty first-line managers never did come on board."

"I was really surprised that many people whom I had labeled as 'hard cases' with a 'bad attitude' did in fact change their behavior dramatically as a result of the training. Of course, some others did not change at all!"

"In my experience, *every* individual has some big dream in their life— something which is their personal top priority. People whom I might have labeled as *not committed* are indeed committed—but to something else. They have reasons for not participating fully, because they have other higher priorities."

"The job proved to be much tougher than I thought."

"I had been highly successful in past jobs and considered myself very knowledgeable about good practice. But I began to realize how little I knew—about quality as a science, about how to create change, about how to work with others in a completely different role with little authority."

"We had put together a detailed technical proposal for management, but when I started the job, my first thought was 'Now what do I do?' It began to dawn on me this was not just a project. The sheer breadth of the challenge became very daunting."

"I had endless surprises, mainly because I knew so little to start with. It took me ages to realize that the consultants I was trusting knew even less than I did. I was surprised at the strength of the resistance to change. I had to learn the hard way that I could get nowhere without the solid commitment of the president. And the job was complex and difficult beyond what I could have imagined at that time."

"Before I started this job, I was already a seasoned senior executive with some big assignments under my belt—like negotiating huge contracts, and leading our efforts in hearings where lawyers cross-examine you for days and the organization's fortunes (and your career) hang in the balance. However, this job is undoubtedly the most challenging I have ever had—and the most fun."

"I was surprised by the credibility given to outsiders. Due to unusual circumstances, I moved back and forward several times between being an internal and an external resource. In each case, within months of joining I had ceased to be an expert in people's eyes, and within months of leaving I was again acknowledged as an expert."

"I was surprised at how the concept of variation changed my thinking—it just kept popping up everywhere I looked. Our problem was that there was too much variation about everything we did. This was a revelation to me."

Hardest things to learn

"I always have difficulty dealing with internal politics—established roles, personalities, personal agendas. No academic background seems to prepare you for this—certainly not a technical one like mine."

"I struggle with my feelings toward those people who relentlessly pursue only their own self-interest. We all act on self-interest, but some people will pursue their own aims with a cynical disregard for the effect on others. I despise this attitude."

"I struggled with the expanding scope of the task."

"My biggest struggle was moving overnight from a background as a technical person to a much higher level of management. And top management's decision did not mean what we had thought—[managers] also thought this was a program."

"I struggled at times with a sense of loneliness and isolation."

"I had to learn to go slow to go fast—I had to learn not to drive others, but to draw them. The person who told me this asked me 'How long did it take *you* to get it?' The answer was, quite a long time! When I let others find their own way, then they begin to go too fast for me at times."

"I struggle to contain my personal impatience for change. Things just don't happen very fast! And I found it hard to accept that there always seemed to be one step back (or more) for every two steps forward."

"I find it hard not to take it personally when things don't work in spite of my best efforts. For example, when I do lots of preparation and have high hopes, but the group session bombs anyway."

"I found it hard to accept that no matter how hard I worked, or how good I was at the job, the outcome was not under my control. If key people were determined not to get on board, or if my boss began to waver, I was just 'pushing on a string.'"

"The harder I would work to compensate for the lack of effort on the part of my boss or others, the more it let them off the hook. I had to learn to stand back and let others take ownership—even if this meant that some things didn't get done."

"Having already had a career as a line manager, I found it hard to get used to acting through others—having responsibility for the outcome, but not owning the resources."

"I struggled with how much *I* had to change, and the need to keep revisiting what I thought I already knew—either to change this or to reaffirm it."

"Someone told us 'When you are learning new things, you have to be prepared to be inelegant.' You will surely make a fool of yourself at times. I hate to look stupid, but I had to accept that I couldn't learn how to create change without paying this price."

"In an operations role, I had believed that I knew all about how my department worked. However, as I began to understand system thinking, I could see gaping holes in my understanding of our organization. This meant that our efforts to create change would go wrong in ways which we could only discover by 'having a go.' System thinking was a very disturbing revelation to me."

"Our organization's functional alignment led to absolute gridlock and despair. It was only when we began to focus on business processes that we started to unlock the grid."

"It was a tremendous struggle for me trying to manage expectations among so many different constituencies, in so many ways. For example, the senior management team expected this to be much like any other program—a plan would be developed, it would be executed, and the job would be done. [Managers] couldn't understand why I couldn't just produce the plan and tell them all what they had to do."

"I had a struggle to hold to a position of refusing to do anything until we had figured out what needed to be done. For example, my colleagues on the senior management team expected a normal plan with very tangible objectives, deliverables, and tasks, and they expected it yesterday. At the same time, they really didn't understand what they wanted the plan to accomplish, or how it could be done—but they believed that they *did* understand."

"I have been working for over a year toward having the formal change agent role transferred to another vice president. This is the right thing to do because his particular line responsibilities give him much more credibility than me with our peers. I finally succeeded recently, but now it is a real struggle for me to give up the formal role—I feel like a mother whose children are leaving home."

The rewards

"I have never had to worry about where my next job would come from or what I would be doing next."

"With this type of experience, one's market value ramps up pretty quickly."

"Although I don't think that this is the primary motivation for most people, there is tremendous career opportunity with this type of work. For some people it is launchpad for their career. There is usually a high level of visibility—across the organization and at different levels. And although you can never seek recognition, it usually comes to you—by association with the overall success of the process."

"I get a kick out of achieving the posted milestones we set for ourselves as a management team."

"I found the personal development of the senior management team very rewarding—seeing this group starting to listen and to address the issues rationally."

"I felt rewarded when I saw the attitudes of my peers change gradually over time—perhaps a 90 degree change of direction the first year, then a bit more, till I could anticipate them being fully on board."

"After feeling for a long time that I was the sole advocate for the process, I found there were many. In board meetings others were starting to speak up and echo what I had been saying. I also welcomed the recognition I received for having led them to this understanding."

"I find it especially rewarding to see people coming on-side—especially those who seemed the most opposed at the outset."

"I feel pleased when people not yet directly involved—perhaps observers on the fringes—stop by my office just to give moral support like 'I'm glad we're doing this,' or 'It's the right thing to do,' or 'Can I do anything to help?' This signals a ground swell of support, and it feels good."

"I find it rewarding to be able to keep quiet and see a group begin to fly— seeing them focus on the customer, and function without intervention from me as the facilitator."

"It may sound stupid, but I got the biggest kick out of seeing meetings begin to work properly—people following agendas, having a timekeeper and a facilitator and so on. This was like the first buds of spring—an indication that something was beginning to happen, that we really could turn this monster around."

"I enjoyed the improved personal relationship with frontline people. When we were on trips together, we used to run out of things to discuss after a few hours of hockey and fishing. This used to be a strain. Now we have fun talking about the things they are doing to get the job done better."

"I love listening to our people explaining their accomplishments to others, their charts and data all over the place, the pride shining from their eyes. They say things like 'I've worked in operations for 20 years, and if you had told me two years ago that we could achieve this level of performance, I would have said you're crazy!'"

"In one of our group meetings a lady in her fifties who had a *terrible* manner in dealing with others spoke up and told us that she realized that some of her ways of doing things were a problem. She acknowledged her unpopularity, and told us she was working on it. I was touched because it was clear that what we were doing was making a difference to her life, because she seemed such an unlikely candidate to respond in this way, and because of the great courage it must have taken for her to speak up like this."

"We offered our housekeeping staff a grade 12 equivalency class. The valedictorian was a 56-year-old immigrant from Eastern Europe whose education had been disrupted by World War II. In her speech she explained how, although the teenagers of today take high school graduation for granted, she had never expected to have this opportunity. She said 'My life-long dream has been to have a grade 12 diploma—you have no idea how I feel.' I still get choked up just thinking about that."

"After I had helped to get the improvement process started across the whole corporation, the staff of my previous department presented me with a memento. The rewarding part for me was the inscription, which indicated that they saw me as their first continuous improvement coach."

"The most rewarding part of this job has been the people I got to work with— the network of other change agents. These are some of the nicest people I have ever met—so open to helping each other. Perhaps it's because we are all in the same boat."

"Working to improve quality within my own line function is always a joy. It is trying to influence others to follow this approach which is sometimes frustrating."

"I found the personal growth tremendously rewarding. I learned more in two years than in the previous 10, about people, about how work gets done, about organizations, about myself. If, like me, you love learning, I can't think of any better situation."

The personal struggles

"Getting the process started is always a struggle—there's so much to do, and it drags and drags before you see the signs that you are winning. It is draining."

"I struggled with the lack of support and understanding by senior management. They are good, caring people in their private lives, but it's not in their business mentality to give a damn about our employees. When costs need to be reduced, they just automatically begin to lay off people—they don't even *think* about how improving key business processes could help to reduce costs."

"I gave myself needless pain by not knowing about support structures or how to access these. I was fortunate to have a few senior colleagues who would let me vent my frustrations and give me moral support—but that was all."

"People don't realize that this is a high-stress job! People who don't understand your role will offer snide comments like 'What did you produce today?' or 'What's it like to have a nice easy number?'"

"At times I wondered if I were going mad—I would alternate between euphoria and deep depression within the space of hours, because I was so used to motivating myself by *tangible* results. I learned to look instead for changes in the attitudes of others."

"I found it hard to sustain my spirit and self-confidence when we were not making much progress. I tended to think 'If only I were better at this . . . or better at that . . . then everything would be working.' I blamed myself for failing to persuade the people that didn't come along. With hindsight I was wrong to take so much on myself. This period of self-doubt was one of the most agonizing times of my life."

"I struggled with the decision whether to compromise when I felt that the process was beginning to falter. I decided instead to keep challenging my boss and peers and confronting them with my concerns—I felt I had that obligation. I'm not certain now that this was the best decision as a professional. Perhaps I could have had more influence if I had bided my time, or tried a different approach. But as an individual I have no regrets—it was time to get out and let someone else have a go. Incidentally, my successor had no better luck than I had with that group of people."

"I always find middle management the hardest group to 'sign up.' For shop-floor people the quality of their work is their main measure of self-worth (especially when standard pay rates are negotiated), and top management can see the benefits of improved efficiency leading to better profits. With middle management I struggled to find the incentives that would bring them on board."

"I had difficulties dealing with people who would talk the talk very convincingly, and then retreat to the familiar—turn around and do something really stupid, thoughtless, and inhumane. I felt I had to confront them with their behavior, and often I did so although this was very risky for my career. I used to struggle to get up the courage to confront them. I wish I had done so more often."

"I struggled with the change in personal relationships with others which I had to accomplish in order to make the new approach work in my own line department—and to be effective as a change agent."

"One of the hardest things for me was the sheer effort required—the number of hours I found I had to put in. This was one of the factors which contributed to a failed marriage."

"I had to learn to cut my losses and walk away from 'hard cases' rather than go on trying to help them change—these individuals had become a challenge, and it is not in my nature to give up!"

"I find it hard to strike the right balance between working on hard and soft issues—it is so easy to end up with an imbalance in one direction or the other. My goal is to find ways of combining these in order to eliminate this problem."

Good decisions

"I had to decide to be less idealistic and to give up unrealistic expectations. Instead, to be effective, I had to recognize what could be done and what could not, and focus efforts where I could make a difference."

"It is hard sometimes to know when it's time to quit, to back off, to save yourself—you need to learn to take such decisions."

"I'm a very confident, self-reliant person. I've worked in quality and as a change agent for decades, and I'm proud of my expertise. So it was soul-wrenching for me to face the realization, during an important project, that I needed to seek counsel. However, I did accept this, and decided to call in another quality expert. The results were an eye-opener. This was the hardest and also the best decision I ever made. The project was, in the end, a stunning success and this changed the course of my professional career."

"My best decision was to open up the scope from a technical project and to tackle the broader issues. It would have been much easier and better for my career not to do this, but it was undoubtedly the right thing to do."

"The best decision I ever made was to leave an organization in order to get out of a situation where I could not win or even achieve anything useful. I should probably have made this decision earlier."

"I decided to go to the CEO and told him 'We must change the way we manage this business. Here's the problem as I see it.' He agreed, and we began a process of developing the mission, vision, values, key processes, and so on. This was the start of a revolution in management practices in our company."

"Until recently I've worked as a change agent from a line function base. During a recent reorganization I was asked to take on one of the 'plum' operational jobs, but decided to turn it down, and to leave operations in favor of a corporate development role which would allow me to devote my efforts to this field. To most people, inside and outside the organization, I appear to have been downgraded. I also realize now that this perception may adversely affect my future career development. I'm not yet sure whether I will regret this decision or not."

"The best and most significant decision I made was to stick with what I believed—about what was the right approach for our organization. It took about 18 months before we could convince *any* of the senior management team—we came at it many times from different angles, and I often thought of giving up. But once we found ways to convince them, it became clear that this approach was right for us."

Decisions regretted

"Poor decisions? I don't know where to start! We've all made so many mistakes. That's how we learn."

"I did not recognize for a long time that I was relying on a sponsor who was really just an advocate. This person just did not have the authority to drive the process."

"I did not realize the need to keep renewing the mandate!"

"Coming into an organization as a fresh face, I quickly saw what was required and laid out a good, sound plan. The senior managers were demanding this plan—it was what they had hired me to do. However, we soon ran into difficulties because it was my plan, not theirs, and we had to start over. With hindsight, I should have resisted the pressure and led my colleagues through a lengthier process, to create a plan which they owned."

"Because of my background, I neglected at first the hard issues such as systems and processes, and pushed exclusively on the soft issues such as leadership and teamwork. As a result we were missing a vital component for the transformation."

"I organized a rollout of mass training before people had a need to do something with it. I learned that it's better to create a need to do things better, *then* offer training to facilitate this."

"I didn't spend enough time learning the 'lay of the land.' For example, I did not realize how important it was in this organization to treat the professional category very differently from others, to avoid insulting and alienating them."

"There were many decisions I regretted! Mainly, I was too dogmatic at first, and preached too much. The grand plan for transforming the organization only meant something to a few people. Rather than try to communicate this, and thus get into debates about the *Grand Plan,* we should have focused attention more on the tangible first steps, and promoted and celebrated the small successes which resulted."

"I don't regret any decisions. I know that in every case I used all the facts available, and drew on the best judgment of those involved. When you have done your best in this way you shouldn't look back or harbor regrets."

Pride of accomplishment

"Winning the Baldrige Award—and being able to share what we had done—was the high spot of my professional career."

"The changes I started have endured—the companies I helped get started are still working on quality, many years later."

"I'm pleased to see what is happening in the public service as a whole today—it is the start of a reawakening. I'm proud that I helped to make this happen."

"There is a division of a company out there which is a living example of my beliefs and theories about how organizations should work. They are enthusiastic and highly successful—they have done everything, they are an exemplar in the industry. This division is like my baby—it's my pride and joy."

"I started a revolution in our company, and I'm proud of that. Very few people in the organization know the role that I played, and I'm proud of that, too. Why? Because that is how the job is supposed to be done—the people involved feel it was *their* achievement, and my main satisfaction lies in seeing it happen. A handful of people—the president and a few others—do understand the contribution I made, and their appreciation and respect also mean a lot to me."

"Above anything else, I am proud of having gained a reputation for great integrity."

"I am proud of the framework for change which I created, and which is now widely used as the central model of our improvement process. This was a breakthrough in helping people to understand."

"I got a great sense of accomplishment when I began to realize that I now knew better than our consultants when it came to decisions about our strategy. We had done our homework and identified first-class consultants. However, on some important issues where our views differed, I began to realize that I had grasped this 'quality stuff' well enough to know better than them what was right for our organization. That was like a coming of age."

"No one thing stands out—but there are many small things. I suppose I'm proud of the number of people I've won over to routinely 'doing the right thing.' For example, we used to be very locally focused in our budgeting—if some travel was required to fix a customer's problem but the travel budget was used up, the customer would suffer. Today we readily move resources around to achieve the best solution for the customer and for the company."

"There's nothing that I feel particularly proud of today—except perhaps that I have learned a lot from my mistakes doing this the first time. I'm 'going round' again so that I *will* have something to be proud of!"

The most important advice

Focusing on the shared goals

"Link all efforts to the organization's objectives. These are the *what*—quality is the *how*."

"Always keep your eye on the goal. In this way you will never be perceived as playing politics or seeking personal gain—you will keep your integrity and credibility, and the trust of your colleagues."

Others' needs, ownership

"Start by understanding the other person's needs and aims—whether they are customers or colleagues."

"Get others involved, give them ownership and help them to succeed."

"Don't try to do it alone! Form alliances with your colleagues, and use your network to access the resources needed."

"Call upon other resources—there is a potentially huge volunteer work force out there of people who will gladly help you."

Winning commitment

"Keep working on your boss till you are sure that he is fully committed. People in these top positions are under tremendous pressure, and are likely to cave in from time to time. You cannot afford to have this happen very often."

"What should you do when the boss does cave in? It all depends upon the people involved and their relationship—you have to decide this for yourself. My style is to confront the issue head on at a suitable moment. I've had bosses say to me months afterwards 'I admire you for telling me what I was doing.' This is a risky approach—he might just decide to get rid of you!"

"When you're satisfied that your boss is fully committed, keep working to sustain this commitment. Never stop!"

Starting small, prioritizing efforts

"My strategy is: First, identify *one* significant, but not impossible, area that needs improvement, and use quality principles to tackle this. When this effort is successful, it becomes easy to convince others. Second, work hard to make other people look good when they are doing the right things. Third, do not try to control everything, since you cannot—the best you can hope for is a kind of 'coordinated chaos' that is moving in the right direction."

"In building support and seeking early small wins, focus your efforts on the areas where there is the greatest enthusiasm. Don't waste unnecessary effort on those who are 'kind of interested' but not committed, nor on those who will clearly never come along."

Flexibility, perspective

"Keep an open mind. Don't get too close so that you cannot see the forest for the trees—step back constantly and try to see the big picture."

"Listen, don't preach; and invite alternative approaches to reach an objective—there are usually many equally good ways to accomplish the goal."

"There is more than one way to skin a cat. Don't try to do things 'by the book.' Every situation is different—and in every situation there are different approaches which will work to produce the same desired outcome."

"Remember that there is no one right way—the best way depends upon the goals and the situation."

Compromise, relationships

"You need to make those compromises which will enable you to continue to have an influence, and to be heard. This might mean deciding not to challenge your colleagues over an issue which might badly undermine your relationship, or where you cannot win. It means being prepared to concede some battles so that the war may be won—saving your ammunition for the vital shots."

"For me it has been so important to achieve a high level of trust and generosity of spirit within the organization. I am known as a 'stirrer,' but people trust my motives, and so my actions are not resented. I have consciously worked to achieve this level of trust by being very open—for example, by sharing with others my personality profile and my management style scores."

Learning
"Understand that this is not a program—it's complex, broad in scope, and full of gray areas, not black and white. You need to be willing to work on *all* aspects, not just those you're already familiar with or most comfortable with."

"Learn as much as you can, as fast as you can."

"Try to gain a real understanding of what a quality approach is all about. This is not easy in our 'act first, think later' North American culture. Then get on with it—but carefully."

"Always be on the lookout for new expertise, for new ideas, for the unexpected 'nuggets.'"

Looking after yourself

Comments on the stresses of the job
"I have no regrets about getting into this type of work—it is a great community with great people, and I've learned so much about all kinds of things. But I'm not sure if this should be a long-term occupation or just a phase of life—it's too demanding, too consuming."

"Having my own line function is a coping mechanism for me—it's like a sand pit I can always go back to when I get frustrated with trying to influence others. It also adds to my credibility. It allows me to experiment with new ideas, and it helps me to be patient with others when it is taking *me* longer than I expected to implement new ideas in my own department."

"For me, this job carries about the same level of stress as the other senior (line management) positions I've had. I think that to manage the stresses it is important to learn not to take things too personally, and yet you cannot become too detached—you need to be a source of enthusiasm for others.

"I like the expression used by someone recently—you need passion, patience, persistence, and pizza. The *pizza* is when you celebrate successes."

"Start off the way you mean to continue. Establish in your mind a few principles you're going to stand for, and walk the talk. Hang on to your vision—when things are not working, keep going back to it and figure out how things *could* work in your vision. Have fun!"

Patience, persistence, perspective, humor

"Don't expect to see any tangible results for a while after you start."

"Patience and persistence are vital. My philosophy is that if someone is opposing the process, it's often because they are irrational. If they're irrational, they usually fail eventually. If I cannot persuade them at first, I just wait—perhaps a year or so—for this to happen, and then leap in with a plan to help straighten things out. Of course this plan involves using a rational approach—a quality approach."

"I accept that others have to go at their own pace. I remind myself that I cannot control everything. When things go wrong, I just say 'CBE—can't be everywhere.'"

"I just ignore how overwhelming the whole job is and keep going. It all works out in the end."

"Once you have satisfied yourself that your message is right, or that you are doing the right thing, don't blame yourself when things don't work out."

"I believe that what ultimately carries the day is the quality of your own personal product—that is, the standards of your own work."

"Don't dump on yourself or you will find that there are many people in the organization who will help you do so—and you can become a scapegoat for everything that's not working."

"It helps me a lot to know that there are always lots of other things I can do— I have very transferable skills—this helps me to make the right decisions and to take personal risks when this is necessary."

"Keep your sense of humor—learn to laugh at yourself, at the whole situation."

"Don't take yourself too seriously!"

Drawing upon moral support

"I always look for a few people whom I trust and who are straight shooters. I use them as confidants and advisors. Mostly they listen and let me talk, sometimes they give advice, occasionally they tell me I'm wrong."

"Develop your personal support system—friends, family, and others who will give you unconditional moral support."

"Find two or three kindred spirits—probably outside the organization—and really use them—as personal coaches, to commiserate, to celebrate. Be prepared to let these people get to know you well personally, to let them see how you really are."

Personal goals and priorities, and celebrating successes

"Try hard to be honest with yourself—about who you are, what you stand for, and what you want to accomplish—and face reality. If you don't do this, you may find yourself pursuing goals which are not the ones you really want. Then everyone loses—you don't end up where you want to be, and your work suffers because your heart isn't in it."

"The job is challenging, so celebrate success whenever you can. When you celebrate, recharge your batteries as much as you can, because there is a lot of charge being drained out and you don't know when you will next be able to get a recharge."

"Know the milestones you want to achieve—particularly the intangible ones. Reach round and give yourself a *big* pat on the back and celebrate when you achieve these milestones."

"I have never been able to figure out a way of getting the job done *and* getting the credit. I wouldn't mind being president, and I've seen others use this type of job as a stepping stone to the top by playing it purely for appearances. When I have had to decide between doing what would be easy and look good, and doing what would get the job done, I always felt compelled to do the latter. Maybe I'm stupid."

Lifestyle

"I decided a long time ago not to become a martyr to this job, even though I believe in it. I've seen other people do this, and lose their families and their personal lives."

"Stay healthy through exercise, good routines, and a balanced lifestyle."

"Squash! Beating the heck out of that little black ball relieves all my frustrations."

"Cryptic crosswords help me shut off."

"Recognize when to stop for a while. Step back, have a complete break like a long vacation, and get it all into perspective."

"Follow some other personal passion or interest which you have control over. Don't end up living just for the job."

LOOKING AFTER YOURSELF

We've discussed at length in this book how you can do a great job for your colleagues and for the organization. Let's now discuss your own needs.

Managing responsibilities

God, give us
grace to accept with serenity the things that cannot be changed,
courage to change the things which should be changed,
and the wisdom to distinguish the one from the other.

—*The Serenity Prayer,* Reinhold Niebuhr

There are two important rules to ensure that you don't drown in a sea of responsibilities that don't belong to you.

1. Only accept responsibility for what you can control or influence. Let the rest go.
2. Don't relieve others of their responsibilities. The change process belongs to the organization, not to you personally. When sponsors go to sleep, try to wake them up. Don't try to compensate by taking over their role—you cannot succeed in this way.

Dealing with emotionally charged situations

You may often find yourself involved in emotionally charged situations which can be trying and exhausting. When major changes are taking place, people at every level are under more stress than normal. Sometimes people need to vent their anger and frustrations, or they may need a sympathetic ear, a confidant, a mediator, or someone to help them think through some perplexing problem.

These situations can be uncomfortable or threatening, and even when the interaction is a positive one (for example, when your valued counsel is being sought) the effort involved is often draining. You can be a great help to your colleagues by being skillful in this type of role—but you need to avoid becoming overburdened in the process.

The skills you need are like those of a counselor, for example,

- Listening carefully and in a nonjudgmental way
- Probing to understand the context and the background, while avoiding personalizing the problem
- Playing back to the person what you are hearing, as a means of clarification and as a step toward possible solutions.

Anyone who wants to can learn these skills, and many people have a natural aptitude for this role. For your own good it is important to recognize that

- Change causes stress and triggers emotional reactions, so some level of flak is inevitable.
- Most of the flak that comes your way is not personal, even though it may feel personal. Mostly people are simply lashing out at the situation. You may receive more than your fair share of concerns—because you symbolize the changes or just because you are there and willing to listen.
- Positive interactions (for example, counseling) with others can be as exhausting as negative ones (for example, fielding complaints). The simple act of caring about the other person's situation is *work*—emotional work—and listening attentively takes concentrated effort.

Valuing yourself and your role

You've probably noticed an insistence throughout this book that you develop the mind-set and the habit of giving away the credit for accomplishments. Ensuring that others receive recognition is an important part of your job, which you cannot fulfill if you are always wondering if you are getting your share.

However, while ensuring that the doers get prompt and full recognition for their efforts, you should not set out to undermine your own worth or the importance of the support role.

- Don't negate or discount other people's positive opinions of you or your work. If a colleague congratulates you on your contribution, acknowledge the compliment and thank him or her. Don't insult your colleague by throwing back the bouquet. And don't indulge in false modesty by claiming that "it was nothing, really." You know how much effort it took!

- Don't agonize in public or parade doubts about your contribution or about the overall direction.

Do request feedback on your efforts and listen carefully. Do reexamine your approach regularly and be open to the possibility that this may fall short in some areas. Do acknowledge others' concerns and be prepared to make changes. Do face any doubts you have about your contribution, and chew these over—in private with your friends. But, forgive yourself for not being superhuman.

Don't volunteer at every opportunity that you "messed up" if you didn't achieve perfection or didn't fully please everyone. And do convey confidence that you know what you're about—and, hence, that the collective efforts of the team are pointed in the right direction. Who is going to trust a navigator who seems anxious and uncertain about his or her own abilities?

- Do strive to ensure that your support role is valued. The top management team has *no hope* of executing a major transformation successfully without the type of support you (and the support network) are there to provide. Separate the person from the position, and feel free to reinforce the value of the role you play.

Rest content that when the process is successful, you will have earned the lasting respect and appreciation of those few key people who understand your role.

Using your personal support network

We previously discussed your *information* network—a wide circle of people, many of whom you may hardly know, but who are willing to help you occasionally with information and technical advice.

In contrast, your *personal support network* is the small circle of people you can count on as friends, to support you, right or wrong. These are the people with whom you will celebrate your successes over a beer, commiserate when, in spite of your best efforts, things have gone wrong, and blow off steam when your level of frustration goes off the meter. It helps if some of them are in a similar role—because they will understand better your successes and your problems—but this is not essential.

Drawing upon the moral support of your personal friends is perhaps the most important single thing you can do to cope with the strains of this or any other stressful job. Use their friendship as a safety net to help you bounce back—to get things back into perspective, laugh at the situation, laugh at yourself. Then you can get back to figuring out your next move.

Don't confuse this type of support with group crying sessions in which people congregate to chant themes like "No one understands," "No one cares about . . ." (quality, service, whatever they do), "There's nothing we can do," and, above all, "Isn't it a shame!" These are rites of failure which legitimize the role of being a victim and reassure those present that there's nothing they can do. This is not your role, and you should never fall into the frame of mind of being a victim, even if you may sometimes be a casualty.

Staying alive

This heading may seem melodramatic, but it's realistic. People in stressful jobs frequently kill themselves or shorten their lives by becoming too engrossed in their work at the expense of a balanced and healthy lifestyle.

There are many sources of advice regarding how to cope with stress—and techniques that cover the gamut of human activity from meditation to diet, sport, and exercise. Here we will offer just a few thoughts.

- Have a life. How would you choose to spend your time if you only had one year left? How much time *do* you have left? Most of us really don't know. Does this help to put the job into perspective?
- Have a body. We tend to overlook the needs of this handy piece of equipment until it begins to complain. An ounce of prevention—exercise, rest, and balanced diet—is better than a ton of transfusions and transplants later. It's hard to express fully one's *joie de vivre* without a functioning body.
- Have a solution. Attend to the underlying causes of stress, as well as using coping mechanisms. The improvement process is falling apart? Get help to figure out why, and tackle the causes that you can influence. You are furious and upset about a colleague's attitude? Confront the situation—or blow off

steam with your friends and let go of your own childish resentment. Squash and meditation may be great to refresh you and to help you bounce back; but, you also need solutions for the problems that are bugging you or undermining your efforts.

Personal growth

Simply performing this work is a major development experience. But it is all too easy to become focused on helping *other* people to learn and to grow, especially when you have acquired a little more knowledge and experience than others in this field.

However, to succeed in the job, and to take advantage of the growth opportunities, *you* need to get into personal growth mode—and stay there. Here's how.

- Remain a humble student. In order to lead a process of learning, you must remain a student yourself. It is good to have confidence in the value of what you have already learned, but you must remain modest about what you still need to learn. You should always be seeking out new information and better ways.

  This isn't just about book knowledge. You should be a student of better ways to do your own job—routinely obtaining feedback from others on your contribution, and seeking to do a better job for them.
- Embrace personal change. It is easy to critique other people's management style, especially when they have been brought up in a different tradition regarding what constitutes good management. It is not so easy to recognize and accept that your own style may fall short by the standards you are promoting. In this respect, you are no different from your colleagues. To find out where you need to change, you must have some reliable, ongoing feedback mechanisms, and take seriously the information these provide.
- Value diversity. It is easy for us to gravitate toward people who are much like us and with whom we feel comfortable—but in doing so we limit our horizons and filter out many learning opportunities. Without even realizing it we may screen out many points of view—those of other professions, of the opposite sex, of visible minorities, of different categories or levels of employee.

Go beyond your comfort level in seeking out others who have very different backgrounds, expertise, and outlooks, and who may not share your views. Understand their opinions and learn from what you discover. When these people are within your organization, try to make them part of the team.

Exhibit 9.3 *A checklist for personal survival and growth.*

Survival

- Be successful in the job—don't take on situations where you cannot succeed!
- Be clear about what you stand for, why you are doing this job, and what you want out of it.
- Adjust your own measures of accomplishment to include the intangible, such as changes in behavior and attitude.
- Only take responsibility for what you can control or influence.
- Learn to be patient and persistent—wait for the best moments for action, then wait for the results to develop.
- Be philosophical—accept that in spite of your best efforts, some things will go wrong. Let these go.
- Learn to accept flak and to listen to concerns without taking it personally.
- Draw upon your friends for moral support and encouragement.
- Have a life—pursue your other personal interests and passions, and keep the job in perspective. Remember that this process is not an end in itself—it is only a means to an end.
- Maintain your sense of humor —or get one quickly!
- Be kind to yourself—forgive your mistakes, celebrate your accomplishments, have fun!

Growth

- Remain a humble student.
- Embrace personal change.
- Draw upon the power of diversity.
- Strive for balance.

Exhibit 9.4 *A role model for quality practitioners.*

"There is much to be learned by studying how Dr. Ishikawa managed to accomplish so much during a single lifetime. In my observation, he did so by applying his natural gifts in an exemplary way: He was dedicated to serving society rather than serving himself. His manner was modest, and this elicited the cooperation of others. He followed his own teachings by securing facts and subjecting them to rigorous analysis. He was completely sincere, and as a result was trusted completely."

—Joseph Juran

- Strive for balance in your learning. Breadth—understanding the full scope of this field—is as important as depth in the critical areas.

Exhibit 9.4 outlines the characteristics of one highly respected person who might be a suitable role model.

KNOWING YOURSELF

There's a rule, I think. You get what you want in life, but not your second choice too.

—*Real People,* Alison Lurie

I hope that this chapter has helped you understand better the nature of the job as others have experienced it.

Is this type of work for you? In trying to answer a question like this, the most difficult task is often to understand ourselves—our own needs and motivations. Why do we do the things we do? This is often a mystery, hidden from us by pride, by self-delusion, or by values and conventions we salute but do not really believe in.

You cannot wholly separate who you are in your job from who you are as a person, and it doesn't make sense to try. The most fulfilled people are those who find a purpose and meaning in their work that matches their personal values and aspirations.

Finding out who you are—at any stage in your life—involves asking basic questions that are difficult to answer, like "What do I stand for?" and "What do I want to accomplish with my life?"

These are not philosophical questions. The answers, if we can find them, light the path to our personal happiness and fulfillment. Our choice of work is a central decision that defines so much of what we are doing with our lives.

What is a good decision for you? It is one that leads toward your primary goal. I believe that this is why people who take on roles like that of change agent (there are many other similar roles) often make decisions that others might view as incomprehensible—or even (mistakenly) as some kind of self-sacrifice.

In my experience, those who choose this kind of work are people who believe strongly in some principles of right and wrong—but they are not do-gooders in search of a feeling of self-righteousness, or crusaders in search of glory. They are people who have decided what is really important to them—like making a difference of some sort—and who are pursuing their personal aims in a fairly single-minded fashion.

When we accomplish our personal goals—such as creating some specific change for the better—we are happy and fulfilled. We are most frustrated when the goal proves to be beyond our reach. The trick is to be informed enough, prudent enough and bold enough to achieve this ultimate goal without losing along the way everything else that's important—personal interests, family, health.

Next Steps

The great end of life is not knowledge but action.

—Aldous Huxley

I hope that you enjoyed the book so far. Perhaps you feel somewhat better informed than when you started. If so, that's great! However, if this was your first real exposure to this subject, there is still a great deal to learn. *This book does no more than scratch the surface of this body of knowledge.* There is a lot more you could, and probably should, read. But reading and studying, however essential, are merely preparation. Real knowledge and the benefits of this come from practical experience of putting these ideas into action.

If you would like to learn more and begin to act (if you are not already doing so), here are some ideas on what to do next.

1. Study the description of the body of knowledge provided in Appendix B, and figure out where you have the least knowledge and experience. It is most important to gain a balanced understanding of this subject and to see it as a whole. So, you should aim to fill in the gaps in your general knowledge before trying to become even more of an expert in the areas you already know well.
2. Begin to develop your own information network, to learn more about what others have done and to find out who you can turn to for reliable guidance.
3. Look for opportunities to test what you have read about—to begin putting the ideas into practice in your work.

If you would like to see change for the better in your organization, there's no need to wait for others. You already know enough to begin to take the initiative, if only to spread these powerful ideas. I hope these will help you to accomplish something that's important to you.

Please give me feedback on the book so that future editions can better meet the needs of other people like you. I have provided a reader response form inside the back cover.

Bon voyage!

Appendices

The following sections are intended as a guide to further learning.

A. Quality in a Nutshell
A summary of the history of the quality revolution, and the characteristics of a quality approach. This is a primer for those who are relatively new to this field.

B. The Body of Knowledge
An overview of the key areas of theory and practice. This is provided to help you identify possible new areas of study.

C. Strategies, Methodologies, Tools, and Techniques
Brief descriptions of some common approaches to improving quality. This section is provided to help you decide whether any of these specific approaches are worth further study in your situation.

D. Sources
A list of recommended sources, including books, materials, and organizations.

A

Quality in a Nutshell

Changing times

Some remarkable changes have taken place in the world during the 1980s and 1990s—some of these, no one would have had the audacity to predict. The mighty Soviet Union crumbled almost overnight. Corporations that seemed invincible, such as General Motors and IBM, have been humbled by the marketplace. The Pacific Rim countries, led by Japan, have become a major economic force in the world.

What do all of these events have in common? They all illustrate how the methods used to manage an organization or a system can determine its fate.

The Soviet Union exemplifies the crippling effects of a common organizational disease: the attempt of a privileged few to run a vast enterprise, largely for their own personal convenience, through a rigid centralized hierarchy, in a command and control style, and often with a cynical disregard for the well-being of stakeholders–whether citizens, employees, or customers.

Many large U.S. corporations have been felled almost as dramatically. These fallen giants have generally been models of a style of working that now seems outdated. Even working harder isn't enough to compensate for the inflexibility and inefficiency of this old style. These same corporations are now working hard—and often with some success—to unlearn some bad habits and assimilate new ways.

At the same time, enterprising and enlightened companies, many of them in resource-poor Pacific Rim countries, have demonstrated that the vast natural wealth from which North American companies benefit (land, water, minerals, energy) can be overshadowed in a competitive battle simply by *working smarter.*

Clearly the old ways, although they may have been an improvement on what went before, and although they helped the Soviet Union and General Motors to the pinnacles of power, are no longer good enough to ensure mere survival, let alone prosperity.

The quality revolution

Throughout history, innovations that provided some military or economic advantage have spread rapidly, and societies that were quickest to adopt the new ways have prospered. Today, a revolution in management philosophy and practices is taking place, making the old methods obsolete. Quality management is at the heart of this change.

Because it is so powerful, this quality approach is being adopted rapidly in leading companies around the globe—not just in the industrially developed world, but in the developing Asian countries whose strategy is to establish industries based upon a highly skilled (and well-paid) work force—rather than remain mere pools of cheap offshore labor. Quality management is being successfully applied today in organizations of all types and sizes, from government agencies and multinational companies to restaurants and retailers; from space agencies and the military to hospitals and schools.

The current popular interest in quality suggests that this revolution in management practices—about fifty years in the making—has gathered momentum. This is seen by some observers as the latest phase of the Industrial Revolution.

In the early 1990s there was much controversy and debate regarding the effectiveness of "total quality management." This was a natural reaction to the overexposure of quality in the media as a trendy management fashion. However, history so far suggests that this revolution is unstoppable. Like Thomas Watt's harnessing of steam power in the late eighteenth century and Ford's mass production methods in the 1920s, it seems inevitable because of the potential for economic and social gain.

Unlike these earlier developments, this revolution is largely invisible to the uninformed observer. Differences in management practices are much less obvious to us than differences in equipment and technology. In our society, management attention is largely focused on what is tangible, material, and explicit. It is hard for us to grasp that the unseen forces unleashed by a quality approach—such as values, relationships, and methods of working—could be as important to the success of our enterprises as technology and automation. As a result, this revolution has been largely overlooked until recently by those not taking part.

What is a quality approach

When we refer to a *quality approach* in this book, we are talking about a way of running an organization. This approach is fundamentally different from the way most organizations operate today. A lot of it seems like common sense—but it is not common practice.

An organization pursuing a quality approach focuses on

- Identifying and satisfying its customers' needs
- Developing and tapping the full potential of its people
- Improving its key processes

These activities are all viewed as prerequisites for success in pursuing other organizational goals.

A quality approach is also based on certain principles, translated into action by a body of proven methodologies, tools, and techniques. Later in this appendix, we will look at definitions of quality and describe the principles of a quality approach. Appendix C describes a few of the fundamental methodologies, tools, and techniques.

Why quality is important

Quality management has created a revolution because it provides a philosophy—and some proven methodologies and tools—for us to discover better ways of getting work done. It helps us organize ourselves better in the work place, harness better our energies and creativity, and serve better those who depend on our efforts.

There is substantial evidence to indicate that a quality approach can improve the ability of a company to survive and to prosper in the face of competition.[7,8] Over time, it can lead to substantial improvements in the quality of products and services, in productivity and the use of assets, in customer retention and market share, and in employee morale and involvement.

The potential benefits are not restricted to private sector companies. A quality approach can also help public sector organizations to fulfil their responsibilities effectively, and at minimum cost to the taxpayer. It can help health care facilities to contain costs while improving service, and provide a process whereby the education system can reexamine and reform itself.

HISTORICAL ROOTS

The history of the quality revolution is a fascinating one, dating back to the 1930s. It features

- A steady evolution in thinking, fostered by various thought leaders and teachers
- The integration of ideas from various countries and industrial sectors
- The spread of this knowledge as its value became more evident

Systems thinking

A modern approach to quality draws on knowledge from many fields. The aim is to create a complete, holistic system of management in which every component of the organization—from the hiring process to the financial systems—is integrated and supports the same philosophy.

This may involve bringing to bear *any* knowledge that can be useful, from any source. However, some integrating principle is required to make disparate ideas and techniques work together. This may be described as *systems thinking.*

Although many people have made important contributions, the roots of the quality revolution are generally traced back to Walter A. Shewhart, of Bell Laboratories, whose ideas launched this system-oriented line of thinking about quality. This part of the quality story is set in manufacturing industries.

In the 1930s, Shewhart published his ideas about how statistics could be used to study variation in the performance of systems. His techniques enable the observer to distinguish between *random* variation—which is inherent in any system—and variation that could be attributed to a specific *cause.*

Understanding variation in this way is the basis for statistical process control (SPC), which enables perfect products to be manufactured by the million without mass inspection. SPC is also invaluable for studying and managing any kind of system, from an industrial machine tool to a municipal school system.

One of Shewhart's students was Dr. W. Edwards Deming, a statistician. During World War II, Deming taught SPC to thousands of American engineers and operators. SPC was in widespread use among suppliers to the war effort, driven by the need for products that would work perfectly every time (such as bullets, which cannot be individually tested after manufacture). SPC was considered so valuable to the war effort that it was classified as secret.

When the war ended, there was a huge worldwide demand for mass-produced goods. Only U.S. industries, untouched by the devastation of the war, could meet this demand. In this booming sellers' market, quality became secondary to sheer volume. The use of SPC—and management concern about quality—declined sharply in the United States.

Deming later mourned that within two years of the end of the war, nothing remained of his work—"not even smoke." Prompted by this bitter experience, he began to develop his own theory of management based upon quality principles. While still based on systems thinking, his theory focuses more on management's responsibilities for quality, rather than the technical aspects. In doing so he also broadens the applicability of this knowledge, as management is a universal activity.

The spread of knowledge between countries

At the end of World War II, the United States offered assistance to the Japanese in rebuilding their shattered country. The Americans provided classes in industrial and scientific management, which whetted the Japanese appetite for knowledge. Soon the Japanese Union of Scientists and Engineers (JUSE) was inviting U.S. quality consultants to teach in Japan. In addition to Deming, they also drew upon the expertise of Juran and Armand V. Feigenbaum, two other leading U.S. consultants in quality management, who provided guidance in the practicalities of implementation.

In 1950, his work discarded by U.S. industry, Deming went to Japan and was paid by the United States, under the Marshall Plan, to teach the Japanese about quality. At that time *Made in Japan* symbolized cheap and shoddy goods, which were hard to sell, even at rock-bottom prices. A few Japanese companies tried Deming's approach, and found—to their surprise—that it worked. They used his theories and philosophy as the foundation for their system of management, which they have been developing and refining ever since.

In 1951, JUSE established the Deming Prize, which became Japan's premier award for industrial excellence. This is presented each year in a nationally televised ceremony. In 1960, Deming was presented with the Second Order of the Sacred Treasure by the prime minister of Japan. Today, he is still honored by the Japanese as one of the main architects of their economic miracle.

It is astounding that while Deming was revered in Japan, he remained almost completely unknown in the West until 1981. It was not until CBS television featured him in a documentary entitled "If Japan can, why can't we?" that the 81-year-old was recognized in North America and able to resume his work in his home country. Deming continued his work until his death at the age of 93 in December 1993.

The Japanese companies that pioneered this approach also incorporated useful ideas from many other sources. In addition to the teachings of Deming, Juran, and Feigenbaum, they studied the leading Western companies of the time, such as IBM and Ford. They used the best of what they found, and then began to develop innovative methods of their own—and their own quality experts and research institutions.

However, a quality approach should not be thought of as a Japanese style of management. Before 1950, the approach to quality in Japanese companies was much like anywhere in the world where modern methods have not been taught—simple, obvious, and disastrously ineffective. Sectors of the Japanese industry—from rice farmers to retailers—that have had less competition, have had less incentive to change, and are still run in traditional ways.

Integration of knowledge from different sectors

Many streams of thought have been brought together to form a more complete approach to quality. Strategies developed in one sector have often proved to be valuable elsewhere.

Manufacturing industry experience

How relevant to other sectors is the manufacturing industry's experience of a quality approach? Is this experience of any value to service companies, the education system, health care, or governments? In the words of Peter Drucker, "Production is not the application of tools to materials—it is the application of logic to work."[9]

Many of the challenges of manufacturing are the same as those faced by all organizations. Most employees of a typical manufacturing company never touch the product. They carry out the universal business tasks of planning, purchasing, marketing, selling, tracking the finances, and managing people.

The manufacture of sophisticated modern products is also an extremely challenging management task. For example, the creation of a modern computer or automobile is a more complex enterprise than can be imagined by those who merely use these devices. Today's standard family automobile contains about 10,000 parts, and takes over a million engineering hours to design. This type of manufacturing bears no relation whatsoever to the common image of a simple, relatively unskilled activity—like "making widgets."

As barriers to international trade have fallen, the manufacturing sector has been the first subjected to true global competition, which has resulted in relentless consolidation.

The weak and the mediocre manufacturers are being pushed against the wall, leaving a few increasingly competent players. These survivors compete for customers—to the benefit of consumers everywhere—and must continue to improve rapidly or perish. This trend is clear in sectors such as automobiles, telecommunications equipment, and consumer electronics.

Not surprisingly, leading manufacturing companies have developed ways of doing business that could be valuable to others. The fiercely competitive environment in which they have been operating is a crucible in which some of the best management practices have been refined. These practices are proving their value in all kinds of settings outside manufacturing—in private sector enterprises of all types and sizes, and in other sectors such as governments and health care.

Service industry experience

Service organizations also have made an important contribution to a quality approach. At the point of contact with the external customer—in any organization—there are special challenges.

When providing a service, the customer is often involved in the delivery process in ways that make the transaction unpredictable. For example, airline passengers influence the time required to board an aircraft by their choices of hand luggage and by how long they take to clear the aisle when settling in their seats.

A service cannot be stockpiled like a product, and, without the buffering effect of stocks, uncontrollable external events can easily cause upsets. For example, if a technical problem forces cancellation of a scheduled flight, that flight window is gone forever, and the subsequent schedule for that aircraft is disrupted. This may affect dozens of subsequent flights—and the connections planned by passengers.

In a service environment, the smooth, predictable flow of work, which is so beneficial in most operations, becomes secondary to satisfying the individual customer at the instant of the transaction. For example, passengers who need help to board the aircraft must be served, even when there are many of them and boarding takes much longer as a result. Furthermore, the customer's perception of a service is heavily influenced by intangibles, such as the feeling that the person providing the service is genuinely concerned for the customer's well-being.

For these reasons, leading companies, whose primary product is service, have developed some special ways of dealing with these particular challenges. For example, a company may empower its people to an unprecedented degree, encouraging them to use their own judgment and initiative to deal with the unexpected. Such empowerment may be based on a value system that promotes trust and cooperative relationships—and guides spontaneous action better than rules and policies.

Leading service organizations are often outstanding in these softer areas. Leading manufacturing companies often excel in the harder areas, such as problem solving and the systematic study of work processes. Outstanding organizations, in any sector, excel by combining strengths in both hard and soft aspects of their management systems.

Quality and the education system

The basic principles of quality management were originally developed in North America. However, by a supreme irony, quality as a field of management science was not taught at all in schools, universities, or business schools in the Western world until the early 1990s—and then only in some pioneering institutions. Why was this?

Like other fields, quality management has its gurus, as well as a foundation of theory set out in management literature. However, because this body of theory cuts across the traditional divisions within educational institutions, educators were slow to recognize its existence. Quality has become accepted and credible only because a quality approach has been shown to work so much better than the methods that were commonly taught—and are still common practice today.

As a result of this educational void, most people in our society, including capable, well-educated people in responsible positions, have little or no understanding of this aspect of management. People immediately recognize terms like *brain surgery, accounting, microelectronics,* or *anthropology* as specialities—fields of valuable, specialized knowledge. However, to most people, the word quality is nothing more than a synonym for *excellence* or *goodness*—the word does not convey the idea of a body of knowledge. Yet there is just as much science in this field as these others. *The first step in gaining access to any field of knowledge is simply to realize that it exists.*

DEFINING QUALITY

It is natural for newcomers to this field to look for some solid foundations on which to build their knowledge—such as clear definitions of the terms used.

When you see the diversity of opinions published about quality in the press, and countless consultants and writers conveying their own distinctive viewpoints, you begin to sense that there is a kind of pillow fight going on in the quality world. There is controversy about what theories are correct, and about what works and what doesn't—as in other rapidly evolving fields, such as atomic physics or genetic engineering. However, there are sound foundations and areas of solid agreement. Here is some basic information upon which knowledgeable and experienced quality specialists agree.

Some basic terms

A good first step is to define the word *quality*. Here are a few definitions culled from various sources.

- "Quality is fitness for use." (Juran)
- "Quality is conformance to requirements." (Crosby)
- "A product or service possesses quality if it helps somebody and enjoys a good and sustainable market." (Deming)[10]
- "Quality means providing our internal and external customers with innovative products and services that fully satisfy their requirements." (Xerox)
- "Quality is what your customer perceives it to be." (Anonymous)

A definition that I like is: "Quality is consistently meeting your customer's needs and expectations, and developing the full potential of the resources used in the process."

These definitions have a lot in common, but why are there so many? Is this a problem? To answer this, let's consider a simple everyday word such as *run*. A typical dictionary may list more than 40 different meanings for *run* in different contexts. For example: my spouse *runs* every morning to keep fit; my nose *runs* when I have hay fever; my yard fence *runs* along the side of the road; my advertisement *runs* in the daily paper.

Like the word run, quality has many widely used meanings and definitions that are appropriate in different contexts. Some are vague or colloquial, others are precise and technical.

The point is that it is a trap to expect or demand a single definition of quality. Let's examine different ways in which the word quality is used (see Exhibit A.1). The word quality is often confused with luxurious or upmarket. Quality practitioners try not to use the word in this way. These are legitimate but casual definitions, which are of little use in managing an organization.

A quality automobile is one that meets the expectations of the customer, whether this individual wants an economical and reliable family runabout, or a luxurious and prestigious sports sedan. Both may be quality vehicles if they meet the needs of their target customers. Burger Heaven may be your ideal choice for a quick snack on a journey, while Chef Antonio's may be your preference for an anniversary. Both establishments provide a quality service if they consistently meet your expectations. You can always clear up any confusion by asking someone what they mean when they use the term *quality*.

The most fundamental distinction between various definitions of quality is the *perspective*. The quality of a specific product or service may be viewed from a customer's perspective, or from the supplier's. Both views are useful, but the customer's view is paramount. One of the most fundamental changes in mind-set that management has to make is to start measuring quality as seen by the customer, not just as it seems from inside the organization.

WHAT IS A QUALITY APPROACH?

We described a *quality approach* as a way of running an organization. Let's expand on this definition.

Just as transportation based on the wheel has taken many different forms—from the rickshaw to the racing car—there are many

Exhibit A.1 *Uses of the word* quality.

Typical phrase	Meaning of the word *quality*	Notes
That is a quality piece of work... car...painting... wine...	Good, excellent, expensive, upmarket, classy, attractive, stylish, high performance.	Normal layman's usage. Subjective—no implication of objectivity, measurement, or precision.
We need to understand the quality levels of our products and services.	Conformance (to specifications, standards, and so on).	Business usage—implies self-measurement against an internal standard. Units of measurement are defects or demerits.
We need to understand how customers view our quality.	Extent of satisfying customer needs, wants, and expectations.	Business usage—implies measurement of customer perception. Units of measurement are proportions of customers expressing certain views; for example, percent satisfied.
We need to improve process quality.	Extent to which the process produces the desired outputs and makes efficient use of resources.	Business usage—measurements typically include process variation, defect rates, inherent levels of waste, requirements for inventory, cycle times.
We must adopt a quality approach to running this organization.	Embodying quality principles and practices.	Business usage—referring to methods used to achieve better outcomes.
Quality has to be considered as important as schedule and cost.	Product quality.	Schedule and cost are quality issues because they impact the customer. In a manufacturing setting, the word *quality* is often used to refer only to the tangible characteristics of the product. This is a trap. The term *product quality* expresses this thought more accurately.

Exhibit A.2 *The dreaded disease* quality-itis.

Here is another way in which the word *quality* is sometimes used.

"We need to employ quality people, doing a quality job, with quality equipment. . . ."

What does this mean? Who knows? This is an example of a syndrome that could be described as *quality-itis*. This highly contagious disease is often observed in the early stages of a quality initiative. In order to convey enthusiasm, and despite a lack of understanding, the word *quality* is inserted several times into every sentence. Apart from this aberration of writing and speech, nothing has changed—everything else is business as usual.

ways of implementing a quality approach. Every organization that starts on the quality journey develops its own unique style of implementation adapted to its own special needs.

However, the main themes and the underlying principles of a quality approach remain constant. A quality approach

- Is led by top management
- Is focused on satisfying customers
- Seeks to involve everyone and to develop human potential
- Is process oriented
- Employs a prevention strategy
- Aims at continuous improvement
- Seeks to create cooperative, win-win relationships
- Is aimed at long-term goals
- Is systematic and methodical
- Is based upon management by fact
- Promotes public responsibility
- Is a holistic approach

It is led by top management

Pursuing quality involves a transformation in management philosophy and methods—not just some minor adjustment. This can be brought about only if senior management is determined to make it happen.

Managers must lead the way through personal example and hands-on involvement. The cultural changes required simply do not take place without such leadership. The commitment of top management to lead the process is the most important requirement for success, and it is often the most difficult to secure and to sustain. This is one of the reasons why a change agent is needed—to ensure that top managers do not forget the reasons why they first made this commitment.

It is focused on satisfying customers

One of the primary aims of the organization is to understand and meet the needs of its customers—even to delight them. Customer satisfaction is often designated as the highest-priority business goal, even though the organization cannot survive unless financial goals also are met. This is not seen as a contradiction—because customer satisfaction is viewed as a prerequisite for good, sustainable financial results—but it does help to convey the message that management thinking has changed.

It seeks to involve everyone, and to develop human potential

The aim is to give everyone—at every level, and in every department—the opportunity to develop and to contribute. A quality approach seeks to harness everyone's energy and creativity in the search for ways to improve.

This includes seeking wide participation in problem solving, planning, and decision making—especially by frontline people who may have no such roles within a traditional management approach. The aim is to put all employees in control of their own efforts and to give them the knowledge, skills, and facilities they need to achieve this self-directed, self-motivated way of working. Like customer satisfaction, this involvement of people and tapping of their creative potential is seen as a prerequisite for the achievement of other business goals.

It is process oriented

Improvement is caused by understanding and changing the system or the process. The organization is a system to be optimized, which is made up of many processes.

Trying to create improvement by focusing only on individuals is fruitless because they work *within* the existing system. The major causes of error and waste are inherent in the system and beyond the control of individual employees. It is management's responsibility to cause improvement by guiding efforts to work on the system.

Similarly, trying to improve by focusing on results alone is fruitless, because the results are determined by the system in use. If the system is not changed, the results will not change. Quality practitioners sometimes offer a nonmedical definition of insanity as "doing things the same old way, and expecting different results."

What is needed is a good understanding of the cause and effect relationships between the process and the results. If this is understood—through study, research, and experiment—then the methods used can be adjusted to improve the outcomes.

This process or system orientation is, therefore, one of the vital, indispensable elements of a quality approach. It provides an elegant, holistic approach to the task of organizing work. The very idea seems inherently superior to trying to manage all the pieces of the system independently: people, equipment, assets, materials, energy, information, policies, and procedures. But best of all, this approach is practical and proven. The methods, tools, and techniques for putting this philosophy into action already exist. This field of knowledge is called *process management.*

It employs a prevention strategy

The strategy for eliminating error and waste is based upon prevention rather than detection. The aim is to eliminate the causes of problems before they occur, rather than simply to detect and rectify shortcomings after the fact.

For example, if a manager simply checks and corrects all reports written by his or her people, this is a *detection* strategy. If he or she works on developing employees' skills so that they can write clear and accurate reports themselves, this is a *prevention* strategy.

Prevention goes hand-in-hand with a process orientation. One of the most effective ways of preventing problems is to study the process or system, in order to eliminate opportunities for error and waste.

It aims at continuous improvement

No matter how much improvement has been accomplished, there are always practical ways of doing even more with less—of providing an even better product or service at the same or lower cost. This seems to run counter to common sense, but it is demonstrated in practice again and again.

One reason for pursuing continuous improvement is that the goalposts are constantly shifting. Customers' needs are always changing. Technology is constantly advancing, thus making possible better ways. Competitors are constantly seeking a new advantage. Putting continuous improvement into practice means discarding forever the old saying, "If it ain't broke, don't fix it."

It seeks to create cooperative, win-win relationships

Management's role changes from cop to coach, and teamwork, rather than internal competition for individual gain, is nurtured and recognized within the organization. Long-term partnerships are also sought with key customers and suppliers. These external relationships may sometimes extend to involvement of key customers and/or suppliers during new product development, sharing of technical expertise and information, or joint quality-improvement projects.

It is aimed at long-term goals

Serious attention is given to developing long-term plans to stay in business. Most organizations have day-to-day operations that demand constant attention to ensure that all is well. However, constantly maximizing short-term results is a sure recipe for disaster, because it prevents investment for the future.

It is systematic and methodical

The foundation of a quality approach is the set of principles and philosophy described here, which may seem nebulous. However, translating these principles into action always involves systematic and planned action using specific methodologies, tools, and techniques. (Some of these will be described later, and Appendix C provides a list.)

For example, to become truly customer-focused, an organization typically has to develop and implement new systems for

- Gathering and analyzing information to determine customer needs
- Translating these into standards and specifications for products and services

- Fielding customer complaints and dealing with these
- Measuring the effect of these efforts on customer satisfaction

Just talking about the importance of customers isn't enough. Systems and methods are required as well as carefully planned actions to implement and fine tune them.

It is based upon management by fact

Decisions are made based upon data and an understanding of the cause and effect mechanisms at work—not simply on the basis of someone's instincts, their position, or their level of authority. This does not mean ignoring soft issues, discounting the judgment of experienced people, or pretending that factors that have not been quantified do not exist.

For example, it may be impossible to measure accurately the extent of losses due to dissatisfied customers—such as lost repeat business, negative referrals, and tarnished reputation. But it doesn't require a rocket scientist to work out that these losses can be substantial and should not be ignored because of a lack of precise data. Approximate data on an important issue is more valuable than precise data on unimportant issues.

Management by fact does mean using data in a methodical way to understand and solve problems. It means obtaining data on important issues, including hard data on soft issues like employee morale and customer satisfaction. It means drawing on the judgment of people with relevant experience (regardless of rank), seeking different perspectives, and separating professional judgment from personal self-interest. Above all, it means facing reality, even when we don't like what we see.

It promotes public responsibility

A quality approach rejects the notion that companies exist only to make a profit, and that their actions, in pursuit of profit, should be constrained only by the law. Rather, a company—or, indeed, any organization—is seen as a part of society, with important responsibilities to its employees, customers, suppliers, and the communities in which it operates, as well as to shareholders.

For a private sector company, profit is essential to survival, but it is not the only measure of success. A focus on short-term profit alone is not a strategy for long-term survival. A successful company aims to give value to its customers, provide satisfying employment, maintain a safe and healthy workplace, preserve the environment, and in doing so make sustainable profits and stay in business. A successful organization is one that serves well all of those who depend upon it, and so benefits our society.

The list of characteristics described so far may seem fairly comprehensive, but there is one more characteristic without which all of the others are almost useless.

It is a holistic approach

The aim is to create a complete system of management in which every component—from the hiring process to the financial systems—is integrated, consistent, and supports the same philosophy.

Picture a senior manager studying the list of quality approach characteristics. For each item, he or she considers whether his or her organization has some applicable program, policy, or system. If there is, a check mark is placed against this item. When the manager is finished, half of the items are checked. "Good, we are halfway there." The problem with this thinking is that *a collection of programs is not a system.*

To be effective, a management system must include a coherent set of values, policies, and mechanisms that are

- Respected and applied throughout the organization
- Integrated into the normal operation of the business
- Applied consistently over time

An effective management system will also take into account both the formal mechanisms used to operate the business and the attitudes and emotions that govern human behavior.

WHAT A QUALITY APPROACH IS NOT

To complete this description of a quality approach, here is a list of common misconceptions. A quality approach is not any of the following.

It is not a project (or a program)
A project has an end point, when one can say "We have accomplished what we set out to do—our task is complete." A quality approach is more like an ongoing journey in pursuit of improvement.

It is not an add-on
A quality approach is not an accessory, like a supercharger that can be bolted onto the side of a sluggish engine to improve its performance. It is more like an effort to redesign and improve the operation of the engine itself.

It is not an employee-motivation program
There is a "stick and carrot" school of employee motivation that seeks to extract the maximum effort from people by treating them like passive creatures who have to be enticed or driven to work. Such programs usually offer no real consideration of employees' needs or concerns, and do not change the relationship between the organization and its employees. These programs may even reinforce poor relationships by communicating implicitly that

- Employees cannot be trusted to do their best.
- Employees are inherently idle and need to be motivated.
- It is the employees' fault if the organization is doing badly—it is employees who need to be fixed.

The major problem with such programs is that, even if they do cause employees to work harder for a while, they have little impact on productivity. The main causes of error, waste, and poor productivity are inherent in the system.

A successful quality approach does result in very high morale—but this is a consequence of efforts to help employees develop their full potential, to involve them in the running of the business, and to harness their ingenuity and talent in the search for improvement. These actions satisfy intrinsic human needs—to take pride in what we do, and to exert some control over our lives.

It is not a marketing ploy
It is not about trying to change the company's image through advertising and public relations, or seeking to convince customers that

this is a great organization when their recent experiences indicate the opposite. It is not about projecting illusions, but rather creating a new reality: products and services that satisfy, or even delight, customers; repeat business; and referrals that flow from this.

It is not a quick fix

A sure-fire technique to capture management attention is to promise instant results. However, experienced and responsible proponents of a quality approach are careful not to raise such false expectations. Rather, they caution that this is a long-term undertaking that requires persistence and hard work.

Clearly, the transformation will not all be done tomorrow. Once they understand, managers may even be overwhelmed at the magnitude of the task. There's no need to be. What is the best way to eat an elephant? One spoonful at a time. Improving quality requires much the same attitude.

It is not a panacea or a guarantee of success

There are many requirements for success in most enterprises—such as managers who understand the business, specialists with various types of expertise, and financial and technological resources. A quality approach can benefit every activity within an organization, but it is not magic.

A quality approach cannot prevent corporate suicide; for example, through lack of insight into technological trends, or lack of consideration of competitive threats. If a company defined buggy whips as its market niche, a quality approach could help it make great buggy whips at a good price and delight its customers who owned buggies. This would not prevent the development of the automobile.

Even with a quality approach, you still cannot operate a legal firm without skilled, knowledgeable, and experienced lawyers. You still cannot run a great restaurant without talented kitchen staff. You will still need to spend millions regularly to keep your semiconductor line up to date. A quality approach can enhance your human and financial resources, but it will not enable you to succeed without them.

There are also many external factors beyond the control of the organization—the cost of capital, labor and materials, legislation, taxation, the state of the economy, and so on. And because it is still impossible to predict the future with precision, luck will always be an important factor. So, a quality approach may be *necessary* for success—even for survival—but it is never *sufficient.*

It is easy to understand, but not easy to do

A quality approach may appear simple—perhaps little more than common sense. But people who have successfully worked on quality for a while will tell you in no uncertain words: "This stuff is *not* easy to do." It is easy to talk about, and play at—but hard to do really well, and even more difficult to sustain the effort over time. The potential benefits are so great that, if it were easy, everyone would be doing it.

There is a huge gap between understanding what is supposed to happen and being able to do it well. After first understanding the rules of golf, years of practice are required before one can finish under par. Understanding the principles and methods is a necessary first step, but proficiency can only be acquired through practice.

Adopting a quality approach involves people practicing new personal work habits and becoming better at

- Remembering to put the customer's interests first
- Considering the larger picture—not just one's own group or department
- Thinking and planning ahead
- Gathering information before making decisions or jumping to conclusions
- Taking time after a problem has been fixed to work on preventing recurrences
- Seeking problem causes in the system, rather than just looking for someone to blame
- Investigating problems rationally and systematically
- Avoiding shooting the messenger when there is bad news that you don't like
- Being consistent in dealings with others
- Burying old grievances
- Listening to others' information and opinions
- Admitting ignorance

- Asking for help
- Giving others credit
- Remembering to say thank you.

So quality is difficult to do because—like raising teenagers, quitting smoking, or mastering the piano—it calls for

- Making a mental effort
- Changing personal behavior
- Exerting self-discipline

It is not dull, mechanical, or boring

Although it is hard work, learning better ways of working together with colleagues, and seeing the tangible results of this, is immensely rewarding for those involved. Here are some of the things that people really enjoy about a quality approach.

- Understanding why you are doing the job this way
- Learning new skills
- Being able to take pride in your work
- Getting positive feedback from customers, rather than complaints
- Having some control over your work environment
- Making things work better
- Reducing daily hassle and waste
- Being part of a team whose members support and value each other
- Being appreciated; being thanked and recognized
- Looking forward to going to work

This approach can and should be fun. Getting deadly *serious* about doing things right doesn't mean becoming deathly *solemn*. Feeling free to have fun is a boost to creativity and innovation—as anyone knows who has taken part in really lively and uninhibited brainstorming. So it should be encouraged. In fact, having fun in the workplace just comes naturally as people learn better teamwork and problem-solving skills and as managers learn to lead without relying on fear.

Working toward measurable goals that you have set for yourself, keeping score, and achieving these goals—this is also enjoyable and satisfying. This is how most games and sports are set up. How many business activities inspire as much enthusiasm and dedication as these silly games that most people strive so hard at without even being paid?

The pursuit of excellence—it's part of human nature. In the words of Robert Townsend: "If you can't do it excellently, don't do it at all. Because if it's not excellent it won't be profitable or fun, and if you're not in business for fun or profit, then what the hell are you doing here?"[11]

B

The Body of Knowledge

This section provides an overview of the key areas of management theory and practice that are particularly relevant to a quality approach. The intent is to help you identify possible new areas of study. Exhibit B.1 maps out some of the key areas of knowledge that you may want to draw upon. Many of the books, videos, articles, and reports you may want to use are related to some part of this map, and we will use this framework later as a way of categorizing these sources of information.

The diagram also illustrates how the body of knowledge is developed over time: by applying practices suggested by theory; by sifting through the evidence of what worked, what didn't work, and why; and adjusting our theories and practices in the light of this information. Exhibit B.1 is not intended to define boundaries or limits—any source of knowledge or information may be valuable. Rather, it describes a basic framework which allows ideas and methods from other sources to be easily integrated and applied in ways that support the organization's objectives. In fact, when we begin to look at best practices, we find methodoloiges such as benchmarking, specifically designed to help us reach out for new knowledge and ideas.

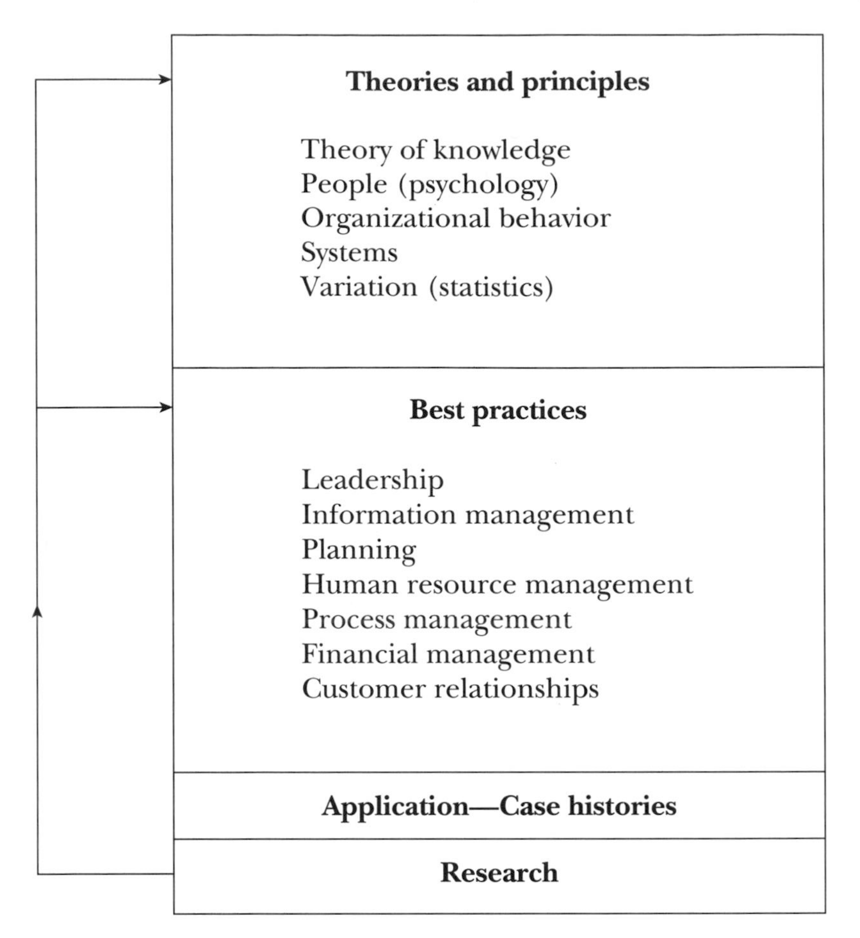

Exhibit B.1 *The body of knowledge.*

Why study theory?

Exhibit B.2 outlines some of the areas of management theory that are key to managing quality. These topics may appear academic, and it is fair to ask "What is the relevance of these theories to getting work done in our organization today?" To answer this, let's consider two examples—variation and psychology.

Variation

When we do not understand variation, we tend to spend much of our time and energy investigating events that are random. We may

Exhibit B.2 *Key areas of management theory.*

Area	Elements
Theory of knowledge	Prediction as the basis of planning and decision making Theory as the basis of prediction, and of further questioning and learning The need for operational definitions as a means of stating facts and to facilitate communication
Theory of variation	The stability, capability, and predictability of processes Uncertainty in measurement Special and common causes The consequences of variation for all types of decision making
Systems theory	The interdependence of all functions, activities, processes The need for an aim that is understood by all The need to optimize the system The idea of loss to society
Psychology	How people interact How people learn How people are motivated How people are affected by change
Organizational behavior	The dynamics of group behavior How organizational structure affects behavior How change occurs in organizations

pay far too much attention to fluctuations in sales figures from one month to the next, or variations in error rates in a department, or short-term movements in stock prices. Month after month we keep asking "what happened?" and looking for causes.

However, when such variations fall within certain limits, you can be fairly certain that they have no specific cause. It is, therefore, a complete waste of time and energy to look for the reasons why. This type of activity is fire fighting at its worst, because there isn't even a fire to be put out.

Once we understand variation, we can learn to quit chasing phantoms. We can begin to examine *significant* chances in performance and trends, and start to focus on understanding, stabilizing, and improving the system. In this way we can find ways to improve the overall level of sales, the average frequency of errors, and so on.

Psychology

Many common management practices ignore what we know about human behavior and motivation, and serve only to create barriers to quality and productivity—fear, frustration, resentment, and conflict. Typical examples of these practices include

- Systems of planning and control that exclude the individuals who perform the work
- Systems of objective setting and appraisal that foster destructive internal competition, conflict, and, hence, suboptimization
- Systems of communication that operate in one direction only (top down), thus blindsiding management and alienating employees
- The tolerance (or even encouragement) of management behaviors that undermine trust, mutual respect, and working relationships
- Attempts to motivate employees, often in ways that insult them, while ignoring chronic problems that undermine their intrinsic motivation

When we understand what psychology has to tell us about human behavior, we begin to recognize the absolute necessity of changing some of our attitudes, behaviors, and systems—if we are to succeed in tapping the energy and the talent of employees.

Exhibit B.3 *Management practices—some subsystems.*

Subsystem	Purpose	Elements
Leadership	Provide direction and purpose	The personal involvement of leaders in communicating the direction, purpose, and operating principles of the organization, and in the planning process.
Information	Capture and communicate information relevant to the running of the organization	Selection of internal and external information required. The capture, processing, and transmission of information. Analysis and use of information to make decisions and solve problems. Measurement of outcomes related to the organization's goals.
Planning	Develop broad goals and translate these into tangible objectives for everyone	Development and deployment of plans at all levels, tracking of progress and outcomes, understanding the relationship between outcomes and the approach.
Human resources	Develop and harness the full potential of human resources	Planning for the development of human resources to meet the organization's goals, including involvement, education and training, reward and recognition, well-being and satisfaction.
Financial resources	Make full use of financial resources	Planning for the financial resources required to accomplish the organization's goals, and projection of the financial outcomes.
Processes	Understand and improve the processes used to perform work	Identifying, studying, measuring, and stablizing processes to meet customer requirements. Improving processes to make better use of resources and to add more value for customers.
Customer focus	Focus attention on activities that have the most value to customers	Identifying current and future customer needs and translating these into requirements for the organization. Managing relationships with customers and measuring customer satisfaction trends and comparisons with other organizations.

These are just two examples of how theory can provide insights that are relevant to getting the job done.

Management practices

Management practices fall naturally into groups representing the common functions found in most organizations. These are the subsystems that most organizations need in order to operate (Exhibit B.3).

C

Strategies, Methodologies, Tools, and Techniques

There are a few strategies, methods, tools, and techniques that you must understand at some level to be effective as a change agent. At a minimum you should know what these approaches are and how they are used, so that you can make informed decisions about which to employ—now and in the future. Some approaches are described here only because you may often hear them mentioned, and it is in your best interests to know a little about them, if only to avoid seeming ill-informed. We will take a brief look at each of the following:

- The seven QC tools
- The seven management tools
- Problem solving
- Process management
- Techniques related to interpersonal relationships and group dynamics
- Hoshin planning (or policy deployment, or management by planning)
- Benchmarking
- Quality function deployment (QFD)

- ISO 9000 (or similar quality system standards)
- Design of experiments (DOE)

The basic QC tools

These are very simple, basic tools for gathering, presenting, and understanding numerical data. They are invaluable for problem solving and for process improvement. They include histograms, cause-and-effect diagrams, check sheets, Pareto diagrams, graphs, control charts, scatter diagrams, flowcharts, and others like forcefield analysis.

This basic toolkit has been expanded to include more than the original seven QC tools—but we've all learned a lot since the term was first coined.

The seven management tools

These are tools for manipulating, presenting, and understanding nonnumerical information—like ideas, or tasks. Sometimes called the *Seven Planning Tools,* they comprise the affinity diagram, interrelationship digraph, tree diagram prioritization matrix, matrix diagram, process decision program chart, and activity network diagram. The names are more intimidating than the tools themselves—none is hard to understand or use. The trick is to find a suitable opportunity to experiment with each—then you will know intuitively how they can help you, and when to use them.

Problem solving

Problem solving in a quality-improvement context calls for data and a search for root causes, which often lie in the work process. This ability to identify process problems is a valuable progression from purely situational problem solving, although both are valid and useful.

Formal approaches to problem solving come in countless different shapes, with different numbers of steps or levels of detail. They generally require the identification of a problem as a starting point. Some approaches also recognize the distinction between an initial quick fix to contain a problem, and the searching out of root causes to prevent recurrences. The seven QC tools are ideal for use within such a problem-solving cycle.

Process management

The term *process management* suggests any formal methodology for studying and analyzing a process and improving it to meet customer requirements and eliminate error and waste. Process improvement is proactive, rather than reactive—it starts with research into the process and into customer (output) requirements. This research generally uncovers many previously unrecognized chronic problems. Usually the process is poorly defined and understood. The first step is to flowchart and define how the current process should work, and strive for consistency in operation. Then the process can be improved. A high-leverage problem is selected and problem-solving methods applied. When this problem has been successfully eliminated, the next most significant problem is tackled, and so on.

Other process management methods

There are many approaches to improving processes. Some focus on reducing cycle time, some on reducing variation, some on reducing costs, and so on. Some focus on individual processes or subprocesses, and some are more concerned with strategy—the deployment of efforts to manage all the key processes in an organization. Two examples follow.

Statistical process control (SPC) is an approach with a strong emphasis on reducing variation, and on measurement as a means of understanding the process and the causes of variation. SPC can be applied to almost any type of system, but is particularly valuable for processes that repeat frequently and/or where historical data have been captured that can be used for statistical analysis.

Process reengineering is a term popularized during the early 1990s. This might be thought of as the big bang approach to process management. Proponents of this approach encourage management to start with a clean slate, rather than the existing process, and to attempt major breakthrough projects. The term also tends to be applied to process-improvement efforts where there is a strong focus on the technological aspects of the process.

Process management is most effective when it is applied within the context of a broader strategy for improving quality. For example, it is vital to consider the human aspects of developing and operating a reliable process—and of running a successful organization. It is also essential to ensure that the customers' needs are considered first and foremost in the redesign. Attempts to reengineer processes while ignoring these other dimensions may be doomed to failure.

It also pays to develop some competence in managing and incrementally improving existing processes before embarking on major high-risk reengineering projects.

Interpersonal and group dynamics

Interpersonal and group dynamics include all the various approaches that may be used to facilitate cooperative behaviors and effective ways of working together. These methods may include reinforcing behavioral ground rules, providing training in interpersonal skills, group dynamics, facilitation, participative management, and so on. Team efforts are a universal feature of organizations trying to improve, and establishing effective meeting habits is often important to the success of these teams.

Problem-solving and process-improvement methods in themselves do change group dynamics, but initiatives based on these hard methods alone can get bogged down or fail due to lack of interpersonal and communication skills within the group. In addition, improved interpersonal and group skills can benefit all of the countless other daily activities that take place in an organization.

Hoshin planning

Hoshin planning is also known as *hoshin kanri, policy deployment,* and *management by planning* (in contrast to management by objectives). This is a comprehensive system for planning, following through, and learning from experience. It involves every part of an organization: first in selecting and defining a small number of key corporate goals; and then in contributing to the accomplishment of these goals.

Hoshin planning differs from other systems of planning in that it makes extensive use of quality management principles and techniques. It evolved from management by objectives, and may be thought of as quality management applied to the process of corporate planning and the implementation of these plans.

The leading Western exponents of this approach include Hewlett-Packard, Ford, Procter & Gamble, Florida Power & Light, Intel, and Xerox. Many of these companies have shared their experiences in the public domain, but literature on this subject only started to become available in English in the early 1990s.

Benchmarking

Benchmarking is the search for the best practices that will lead to superior performance in some business activity. Benchmarking involves systematically studying how others tackle some specific process, and what levels of performance they achieve. This information enables more challenging goals to be set, and provides valuable insights and ideas for accomplishing these goals.

The word benchmarking is sometimes used to denote comparison of levels of performance only, rather than comparison of methods. To avoid such confusion, it may be useful to draw a distinction between *process* and *results* benchmarking. Results benchmarking on its own is of limited value because it does not provide any insight into the reasons for performance differences, nor ideas for improvement.

Quality function deployment

QFD is a way of making the voice of the customer heard throughout an organization. QFD is a systematic process for capturing requirements in the customer's language and translating these into precise, measurable requirements for use within the company. The result is a set of priorities and target values for the design of products and services. The translation process may be repeated to provide requirements that must be met throughout the supply chain—production, and even suppliers.

ISO 9000 standards

ISO 9000 is a series of international standards that define the elements of a quality assurance system, focusing on the procedures and documentation required to demonstrate conformance to specifications for the purposes of a contract. These standards have evolved from the military procurement standards of the 1980s.

ISO 9000 standards have come to be widely used as a basis for formal registration schemes whereby organizations can submit themselves to a periodic audit by an accredited registrar, and thus become listed in a published register.

These standards can be valuable as the basis for efforts to improve documentation and to establish some process discipline—if this is part of a broader strategy for improving quality.

ISO 9000 standards do not address issues such as the human dimension or relationships with customers. However, their most significant limitation is that they call for evidence of a system, but do not call explicitly for evidence that the system is effective in terms of end results. Purchasing from an ISO 9000-registered supplier does provide some assurance that the supplier is using a documented process. However, this does not in itself provide assurance that the product or service provided will be satisfactory to the customer.

Design of experiments or Taguchi methods

DOE or Taguchi methods are statistical techniques for revealing how variations in process parameters affect the final product.

DOE is used to understand how variation in the production process (for example, temperature, proportions of ingredients, dimensions) affect the characteristics of the finished product. This understanding enables products to be designed to be robust; that is, insensitive to variation in those production parameters that are hard to control, and therefore easy to produce consistently.

DOE also identifies which process parameters must be controlled tightly to obtain consistent results. These parameters can then be monitored using SPC techniques. DOE is therefore a natural partner for SPC.

DOE is a misfit in this list of approaches because it is not universally applicable. It is used to assist in the design and production of tangible products, where the laws of physics and chemistry are paramount. It is not applicable to administrative or management processes.

These are by no means the only methods used in implementing a quality approach. Virtually any existing system or technique may be applied, adapted, or built upon. For example, a quality approach always involves gathering better information about customers and employees—about their needs, opinions, and concerns. The common techniques used to accomplish this, such as surveys and focus groups, are well-developed expertise that should be tapped. What usually needs to change is the type of information sought and the methods used to respond to the findings.

D

Sources

The purpose of this section is to identify some of the most valuable information sources—books, videos, associations, and so on—and to help you decide where to concentrate your learning efforts.

BOOKS AND VIDEOS

The following (Exhibit D.1) are a few sources that will allow you, with limited effort, to glean much of what you may need. You may have considerable depth of expertise in some of the areas we have mentioned, but you may find it more valuable to start filling in the voids in your broader knowledge than to become even more expert in your specialty.

For example, no one should undertake this type of work without a basic understanding of how organizational change occurs, and how individuals cope with change. However, this is not a difficult field to understand, and just a few hours of study—in a class, or with a book or a video—may provide you with most of the insight you need to get started.

These are all useful books. For those areas that you choose to study further, you will not go wrong by picking any that are listed.

Exhibit D.1 *Sources of information—Educational material.*

Overviews of a quality approach Various perspectives on the big picture	Aguayo, Rachel. *Dr. Deming: The American Who Taught the Japanese about Quality.* St. Louis: Fireside, 1991. Albrecht, Karl, and Ron Zemke. *Service America.* Homewood, Ill: Dow Jones-Irwin, 1985. Berry, Thomas H. *Managing the Total Quality Transformation.* New York: McGraw-Hill, 1991. Clemmer, Jim, Barry Sheehy, and Achieve Associates. *Firing on All Cylinders.* Toronto: Macmillan of Canada, 1990. Crosby, Philip B. *Quality Without Tears.* New York: McGraw-Hill, 1984. Davidow, William H., and Bro Uttal. *Total Customer Service.* New York: Harper Perennial, Harper Collins, 1989. Deming, W. Edwards. *Out of the Crisis.* Cambridge, Mass: Massachusetts Institute of Technology, 1990. Feigenbaum, Armand V. *Total Quality Control.* New York: McGraw-Hill, 1991. Garvin, David A. *Managing Quality.* New York: Free Press, 1988. Green, Richard Tabor. *Global Quality: A Synthesis of the World's Best Management Methods.* Homewood, Ill.: Business One Irwin, 1993. Harrington, H. James. *The Improvement Process.* New York: McGraw-Hill, 1987. Imai, Masaaki. *Kaizen.* New York: Random House, 1986. Ishikawa, Kaoru. *What Is Total Quality Control.* Translated by David J. Lu. Englewood Cliffs, N.J.: Prentice Hall, 1985. Juran, Joseph M. *Juran on Leadership for Quality.* New York: Free Press, 1989. McCloskey, Larry, and Dennis Collet. *TQM: A Basic Text.* Methuen, Mass.: Goal/QPC, 1993. Scherkenbach, William. *The Deming Route to Quality and Productivity.* Washington, D.C.: CeePress Books, 1986. Walton, Mary. *The Deming Management Method.* New York: Perigee, 1989. The Deming Library (video), produced by Films, Inc., Chicago, 1987–93.

Exhibit D.1 *(Continued)*

Case studies Stories about the renewal of organizations, using quality principles and practices (or a similar approach), or about the consequences of failure to change	Berwick, Donald M., A. Blanton Godfrey, and Jane Roessner. *Curing Healthcare.* San Francisco: Jossey-Bass, 1990. Carlzon, Jan. *Moments of Truth.* Cambridge, Mass.: Ballinger Publishing, 1987. Hale, Roger L., Douglas R. Hoelscher, and Ronald E. Kowal. *Quest for Quality.* Exeter, N.H.: Monochrome, 1993. Hudiberg, John J. *Winning with Quality: the FPL Story.* White Plains, N.Y.: Quality Resources, 1991. Kearns, David T., and David A. Nadler. *Prophets in the Dark.* New York: Harper Business, 1992. Keller, Maryann. *Rude Awakening.* New York: Morrow, 1989. Osborne, David, and Ted Gaebler. *Reinventing Government;* New York: Plume, Nal-Dutton, 1993. Petersen, Donald E., and John Hillkirk. *A Better Idea: Redefining the Way American Companies Work.* Boston: Houghton Mifflin Company, 1991. Rayner, Bruce. *Trial-By-Fire Transformation.* Boston: Harvard Business Review, 1992. Sewell, Carl, and Paul B. Brown. *Customers for Life.* New York: Doubleday, 1991. Tichy, Noel M., and Stratford Sherman. *Control Your Destiny or Someone Else Will.* New York: Doubleday, 1993. Watson, Gregory H. *A World of Quality: The Timeless Passport.* Rochester, N.Y.: XQS Press, Xerox Corporation, 1992. Walton, Mary. *Deming Management at Work.* New York: G.P. Putnam's, 1990.
Evidence of the benefits of improving quality	Buzzell, Robert D., and Bradley T. Gale. *The PIMS Principles: Linking Strategy to Performance.* New York: Free Press, 1987. GAO study. *Management Practices.* Washington, D.C.: General Accounting Office, 1991. Womack, James P., Daniel T. Jones, and Daniel Roos. *The Machine that Changed the World.* New York: Rawson Associates, 1990.

Exhibit D.1 *(Continued)*

Assessment and awards	Brown, Mark Graham. *Baldrige Award Winning Quality.* 3d ed., White Plains, N.Y.: Quality Resources and Milwaukee, Wis.: ASQC Quality Press, 1993. Hart, Christopher W.L., and Christopher E. Bogan. *The Baldrige.* New York: McGraw-Hill, 1992. Award criteria can be obtained from the administrative bodies listed in Exhibit D.2.
Leadership, management, ethics	Block, Peter. *The Empowered Manager: Positive Political Skills at Work.* San Francisco: Jossey-Bass, 1987. Brandt, Steven C. *Entrepreneuring in Established Companies.* Homewood, Ill.: Business One Irwin, 1985. Covey, Stephen R. *The Seven Habits of Highly Effective People.* New York: Fireside, Simon & Schuster, 1990. Covey, Stephen R. *Principle-Centered Leadership.* New York: Fireside, Simon & Schuster, 1991. Depree, Max. *Leadership Is an Art.* New York: Doubleday & Co., 1989. Kotter, John P. *Force for Change: How Leadership Differs from Management.* New York: Free Press, 1990. Kouzes, James M., and Barry Z. Posner. *The Leadership Challenge.* San Francisco: Jossey-Bass, 1987.
Planning	Akao, Yoji. *Hoshin Kanri.* Cambridge, Mass: Productivity Press, 1991. Barker, Joel Arthur. *Discovering the Future—the Business of Paradigms.* Burnsville, Minn:, Charthouse, 1988. ———. *Discovering the Future—the Power of Vision* (video). Burnsville, Minn.: Charthouse, 1990. King, Bob. *Hoshin Planning.* Goal/QPC research report. Methuen, Mass.: Goal/QPC, 1989. *Total Quality Management Master Plan.* Goal/QPC research report. Methuen, Mass.: Goal/QPC, 1990.
Customer satisfaction	Hauser, John R., and Don Clausing. *The House of Quality.* Boston: Harvard Business Review, 1988. Heskett, James L., W. Earl Sasser, Jr., and Christopher W. L. Hart. *Service Breakthroughs: Changing the Rules of the Game.* New York: Free Press, 1990. *Quality Function Deployment.* Goal/QPC research report. Methuen, Mass.: Goal/QPC, 1989. Whiteley, Richard C. *The Customer Driven Company.* Redding, Mass.: Addison-Wesley, 1991. Zeithanl. Valeria A., A. Parasuraman, and Leonard L. Berry. *Delivering Quality Service: Balancing Customer Perceptions and Expectations.* New York: The Free Press, 1990.

Exhibit D.1 *(Continued)*

Systems and processes	*The Customer Is Always Dwight* (video). Chicago: Video Arts, 1989. Davenport, Thomas H. *Process Innovation: Re-engineering Work through Information Technology.* Boston: Harvard Business, 1992. Hammer, Michael, and James Champy. *Re-engineering the Corporation.* New York: Harper Business, 1993. Harrington, H. James. *Business Process Improvement.* New York: McGraw-Hill, 1991. Jablonski, Joseph R. *Implementing Total Quality Management.* Albuquerque, N.Mex.: Technical Management Consortion, 1992. Juran, Joseph M. *Juran on Planning for Quality.* New York: Free Press, 1987. Rummler, Gary A., and Alan P. Brache. *Improving Performance: How to Manage the White Space on the Organization Chart.* San Francisco: Jossey-Bass, 1990. Senge, Peter M. *The Fifth Discipline.* New York: Doubleday, 1990.
Human resource management	Bolt, James F. *Executive Development: A Strategy for Corporate Competitiveness.* New York: Harper Business, 1989. Byham, William C., and Jeff Cox. *Zapp! The Lightning of Empowerment.* New York: Fawcett Columbine, 1988. Kohn, Alfie. *No Contest: The Case Against Competition.* Boston: Houghton Mifflin Company, 1986. Ryan, Kathleen D., and Daniel K. Oestrich. *Driving Fear out of the Workplace.* San Francisco: Jossey-Bass, 1993.
Organizational dynamics, culture	Beer, Michael, Russell A. Eisenstat, and Bert Spector. *The Critical Path to Corporate Renewal.* Boston: Harvard Business School Press, 1991. Hofstede, Geert. *Cultures and Organizations: Software of the Mind.* Berkshire, United Kingdom: McGraw-Hill Europe, 1991. Mintzberg, Henry. *Mintzberg on Management.* New York: Free Press, 1989.

Exhibit D.1 *(Continued)*

Teams, facilitation	Katzenback, John R., and Douglas K. Smith. *The Wisdom of Teams: Creating the High-Performance Organization.* Boston: Harvard Business School Press, 1993. Kayser, Thomas A. *Mining Group Gold: How to Cash in on the Collaborative Brain Power of a Group.* El Segundo, Calif: Serif Publishing, 1990. Scholtes, Peter R., et al. *The Team Handbook.* Madison, Wisc.: Joiner Associates, 1988.
Consulting	Bellman, Geoffrey M. *The Consultant's Calling: Bringing Who You Are To What You Do.* San Francisco: Jossey-Bass, 1990. Block, Peter. *Flawless Consulting.* San Diego, Calif.: Pfeiffer & Co., 1981. Scheim, Edgar H. *Process Consultation,* Vol. I. Redding, Mass.: Addison-Wesley, 1988.
Managing change	Bridges, William. *Transitions.* Don Mills, ON: Addison-Wesley, 1980. Conner, Daryl R. *Managing at the Speed of Change.* New York: Villard Books, Random House, 1993. Kanter, Rosabeth M. *The Change Masters.* New York: Touchstone Books, Simon & Schuster Trade, 1985. Woodward, Harry, and Steve Bucholz. *Aftershock—Helping People through Corporate Change.* New York: John Wiley & Sons., 1987.
Community-based initiatives	Schwarz, Robert A. *Midland City: Recovering Prosperity through Quality.* Milwaukee: ASQC Quality Press, 1989. Weisbord, Marion. *Productive Workplaces: Organizing and Managing for Dignity, Meaning and Community.* San Francisco: Jossey-Bass, 1987.

Exhibit D.1 *(Continued)*

Other strategies, methodologies, tools, and techniques	Balm, Gerald J. *Benchmarking.* Schaumburg, Ill.: QPMA Press, 1992. Brassard, Michael. *The Memory Jogger.* Methuen, Mass.: Goal/QPC, 1988. ———. *The Memory Jogger Plus.* Methuen, Mass.: Goal/QPC, 1989. Camp, Robert C. *Benchmarking.* Milwaukee: ASQC Quality Press, 1989. Ishikawa, Kaoru. *Guide to Quality Control.* White Plains, N.Y.: Quality Resources, 1986. Jura, Joseph M., and Frank M. Gryna. *Juran's Quality Control Handbook.* New York: McGraw-Hill, 1988. Kume, Hitoshi. *Statistical Methods for Quality Improvement.* New York: Quality Press, 1987. Stalk, George, Jr., and Thomas M. Hout. *Competing Against Time: How Time-Based Competition is Reshaping Global Markets.* New York: Free Press, 1990.

ORGANIZATIONS

Exhibit D.2 *Sources of information—Organizations.*

American Society for Quality Control P.O. Box 3005 611 East Wisconsin Ave. Milwaukee, WI 53201-3005 USA Tel.: 800-248-1946 Fax : 414-272-1734	American Productivity & Quality Center and Benchmarking Clearing House 123 N. Post Oak Ln. Houston, TX 77024 USA Tel.: 713-681-4020 Fax : 713-681-8578
National Institute for Standards and Technology (NIST) Route 270 and Quince Orchard Rd. Administration Bldg., Room A537 Gaithersburg, MD 20899-0001 USA Tel.: 301-975-2036 Fax : 301-948-3716	Federal Quality Institute PO Box 99 401 F Street N.W. Suite 231 Washington, DC 20044-0099 USA Tel.: 202-376-3747 Fax : 202-376-3765
Japanese Union of Scientists and Engineers (JUSE) 5-10-11, Sendagaya, Shibuya-Ku Tokyo 151 Japan Tel.: 003-352-2231 Fax : 003-356-1798	National Quality Institute Tower 1, Constitution Square 360 Albert St. Ottawa, ON K1R 7X7 Tel.: 613-526-3384 Fax : 613-526-5659
The European Foundation for Quality Management Building "Reaal" Fellenoord 47a 5612 AA Eindhoven The Netherlands Tel.: +31 40 461075 Fax : +31 40 432005	American Management Association P.O. Box 319 1 Trudeau Rd. Saranac Lake, NY 12983 USA Tel.: 518-891-1500 Fax : 518-891-0368

Notes

1. Research cited by Daryl R. Conner in his keynote speech to the Computer-Based Training Conference, April 3, 1988.

2. Robert Fritz, *The Path of Least Resistance* (New York: Ballantine Books, and Toronto: Random House, 1989).

3. Joel Barker, *Discovering the future: the power of vision* (videotape) (Minneapolis, Minn.: Charthouse, 1988).

4. Holmes-Rahe Scale of stress ratings. Thomas Holmes and Richard Rahe, *The Journal of Psychosomatic Research* 11 (1967): 213–18.

5. Elisabeth Kübler-Ross, *On Death and Dying* (London: Macmillan, 1969).

6. William Bridges, *Transitions—Making Sense of Life's Changes* (Don Mills, ON: Addison-Wesley, 1980).

7. Robert D. Buzzell and Bradley T. Gale, *The PIMS Principles: Linking Strategy to Performance* (New York: Free Press, 1987).

8. U.S. GAO, *Management Practices: U.S. Companies Improve Performance through Quality Efforts,* report GAO/NSIAD-91-190 (Washington, D.C.: United States General Accounting Office, 1991).

9. Peter F. Drucker, *The Practice of Management* (New York: Harper and Row, 1967), 122.

10. W. Edwards Deming, *The New Economics* (Cambridge, Mass.: Massachusetts Institute of Technology, 1993), 2.

11. Robert Townsend, *Up the Organization* (New York: Knopf, 1970), 54.

Index